The BACKYARD HOMESTEADER

28 27 26   7 6 5

Previously published as *Country Skills: A Practical Guide to Self-Sufficiency.*

This edition published in the United States of America in 2024 by
Gibbs Smith
P.O. Box 667
Layton, Utah 84041

The authorized representative in the EEA is Simon and Schuster Netherlands BV, Herculesplenin 96 3584 AA Utrecht, Netherlands, info@simonandschuster.nl

www.gibbs-smith.com

This book was conceived, edited, and designed by Quarto Publishing, an imprint of The Quarto Group
1 Triptych Place,
London SE1 9SH

E TSSH

Designer: Louise Turpin
Managing Editor: Julie Brooke
Assistant Editor: Jo Morley
Production Manager: Rohana Yusof
Publisher: Samantha Warrington

Printed and bound in China

Gibbs Smith books are printed on either recycled, 100% post-consumer waste, FSC-certified papers or on paper produced from sustainable PEFC-certified forest/controlled wood source. Learn more at www.pefc.org.

Library of Congress Cataloging-in-Publication Data: 2020942769

ISBN: 978-1-4236-5678-4
E-ISBN: 978-1-4236-5679-1

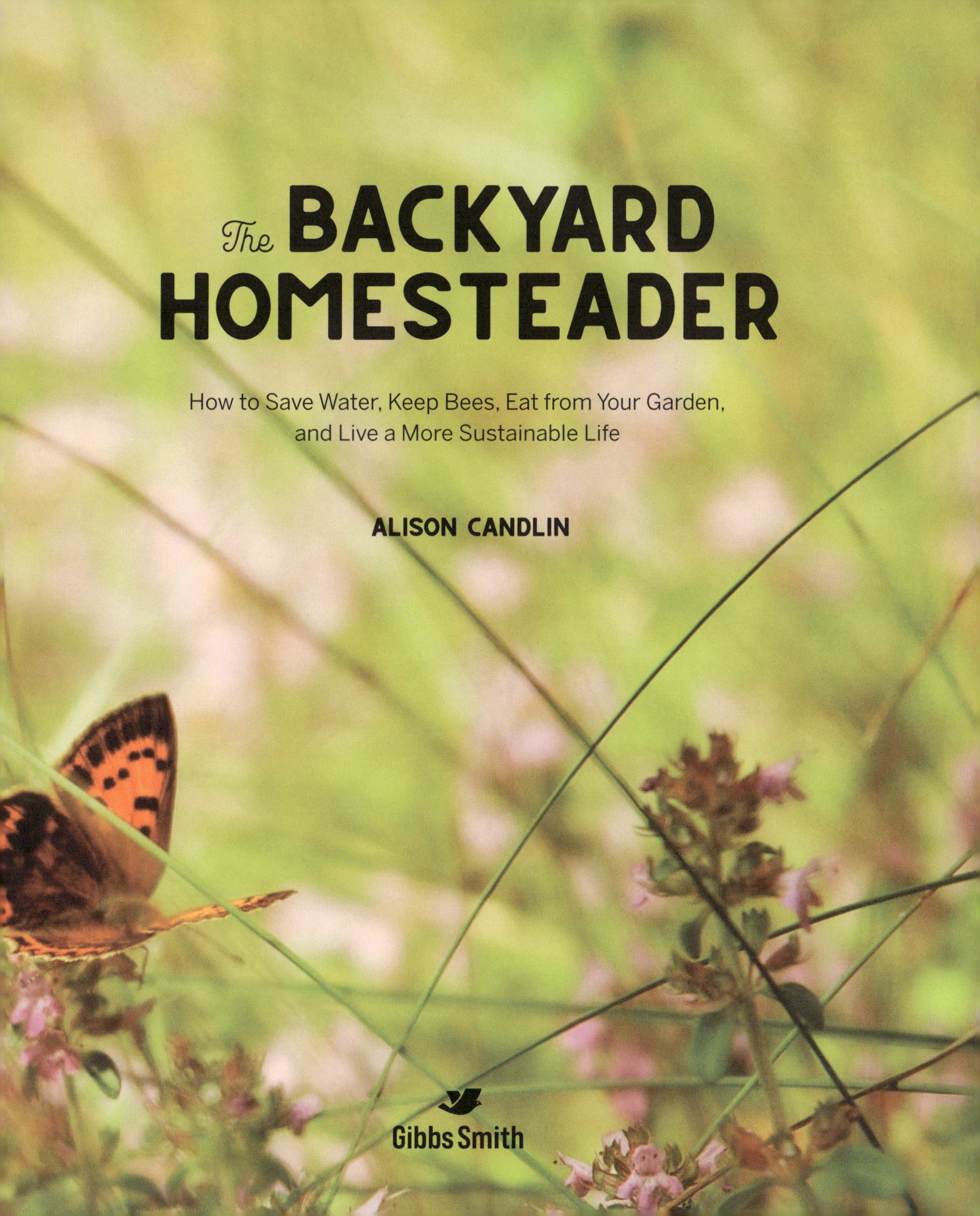

The BACKYARD HOMESTEADER

How to Save Water, Keep Bees, Eat from Your Garden, and Live a More Sustainable Life

ALISON CANDLIN

Gibbs Smith

Contents

Directory of Pests and Diseases

Keeping Animals

Food from Nature

Preserving Your Produce

Water and Energy Conservation

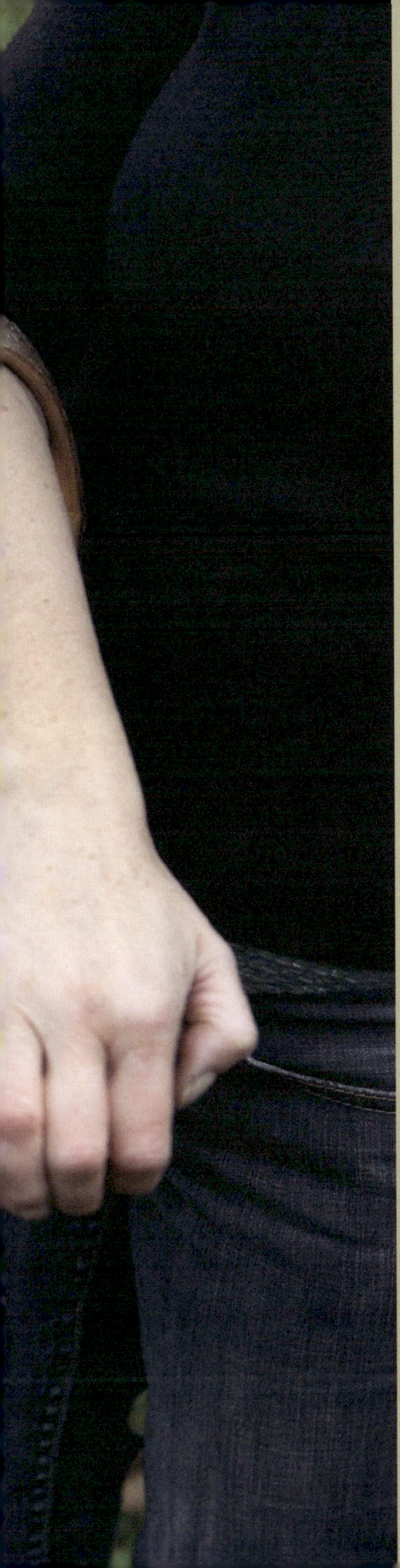

FOREWORD

This book is for those who want to approach the "good life" gently and gradually. Perhaps you want to try your hand at growing some fruit and vegetables, producing your own honey or eating your own eggs, drinking milk or eating cheese from your goats, or rearing your own pigs so you can enjoy home-produced meat. It is for those people, too, who would like to make more use of the free produce that may be found beyond the garden—in the fields, forests, and rivers, or by the coast.

In time, you may want to turn your hobbies into a way of life and aim to become fully self-sufficient, but it is best to start slowly and get a feel for whether the homesteader's life is right for you. Do not try to grow all your own fruit and vegetables, and keep all the animals discussed in this book in your first year. Instead, tackle it little by little, finding what you most enjoy doing, and giving yourself time to learn from your mistakes.

YOUR PROPERTY

You need very little land to produce at least some of your own food. A tiny backyard—even a deck or a patio—can yield a surprising amount of produce if you use it efficiently.

However big your property, always try to put back into your soil what you take out of it. No ground will go on producing crops or supporting livestock of any type unless it is well maintained. This is not as daunting as it may sound and this book will help you to understand the needs of the soil and the effects of growing various crops so that you can ensure a good balance.

NEIGHBORS AND OFFICIALS

The relationships you establish with those who live close by can make a surprising difference to the success of your operation.

People who have lived in the area for some time often have valuable hints and advice on what does and does not grow well, and knowing this can save you a great deal of time, money, and exasperation. In return, be mindful of your neighbors when deciding where to keep your livestock. Few people will thank you for shattering their early morning peace with a crowing rooster, but a little polite conversation and the odd box of fresh fruit will go a long way towards building a harmonious relationship.

If you intend to start a homestead with livestock and a fairly large cultivated area, you will often find that local farmers will be willing to offer advice and help—and maybe even share equipment.

Essential rules and regulations

Before you start, contact the government agencies that exist to offer advice to homesteaders and small farmers. Start with your local Cooperative Extension Office. They will be able to guide you in planning, getting the appropriate licenses for keeping livestock, choosing crops to grow, and give you information about pests and diseases that could be a problem. There are rules and regulations that apply to the keeping of almost all livestock, and in some areas you are not allowed to keep such animals at all.

STORAGE SPACE AND OUTBUILDINGS

The amount of storage space and outbuildings you will need will depend on the scale and scope of your home production plans.

Even if your garden is confined to your backyard, you are unlikely to be able to eat all the vegetables and fruit you grow as you harvest them. And, in any case, the whole point of self-sufficiency is to be able to enjoy your produce all year round.

Root vegetables can be stored in their natural state (*see* pages 207–208), but for this you either need to allocate them a patch of ground (for trenching), or some sort of weather- and rodent-proof shed.

Keeping preserves and other food

You will need ample shelf space in a cool, well-ventilated, but shady pantry for storing chutney, pickles, sauces, jams, and other preserved and canned foods. And a deep freezer is another essential item—if you can, invest in a chest freezer that can be kept as close as possible to the kitchen.

Making space for livestock

You will find out what you need to house livestock in the pages of this book devoted to each type of animal (*see* pages 152–177), but again there are further space considerations. If you intend to keep goats, you should give some thought to the milking conditions and whether you can accommodate some sort of dairy (goats should not be milked in the shed in which they are kept, because conditions are not hygienic enough).

You will also need room to keep eggs as they come in, storing them in such a way that they can be used in strict rotation. Eggboxes can take up quite a lot of room on a shelf or in a pantry, yet they must be kept somewhere cool, where there is no danger of the eggs being smashed.

Storing animal feed

It is most economical to buy bulk feed for your animals, but you must weigh the savings against the storage room you will need. Hay and straw are bulky and must be kept in a weatherproof shed. Other more concentrated feeds must be kept in rodent-proof containers, and these too should be housed in a shed or outbuilding of some sort. Anything rodent-proof must be made of galvanized metal, ideally, with a lid that fits right over the container. Failing this, a heavy, thick wooden barrel will keep the pests out for some time but plastic presents no challenge at all.

The closer your storage buildings are to the house or to the animals' quarters, the more convenient you will find them, but always make use of any old shed or outbuilding that is already there on your property before spending money on new buildings. Don't forget about the attic, basement, or garage, and make full use of this added storage space, erecting a system of shelves if there is enough room.

CONSERVATION

Conservation of energy and the environment are often high on the list of priorities for people striving for self-sufficiency. These concerns go hand in hand with a wish to minimize food miles, grow and produce food in organic or ethically sound ways, and to move away from a wasteful, disposable lifestyle.

Being aware of ways to reduce energy and water consumption and doing all you can to collect and recycle water around the garden and your fruit and vegetable beds all help to conserve the environment. On top of this, technology is improving all the time and making it more and more feasible to generate your own energy with solar panels, wind turbines, or water turbines if you are lucky enough to have a stream or river. You may even be able to sell some excess power back to the grid (*see* page 246).

Generating and conserving energy

Solar technology has come a long way in the last decade, but for homesteaders in rainy and overcast areas, they are still of limited use. In other areas, solar systems are likely to pay for themselves in savings within a couple of years. Federal, state, or provincial programs may offer incentives to help towards the installation of energy-generating systems, as well as energy conservation measures, such as upgrading household insulation. Contact your utility company and a local sustainable energy association to see whether you qualify for some kind of financial incentive or homeowner grant.

Sustainable energy sources

If you have a good supply of sustainable wood on your property, wood-burning systems are a good option for heating your home and hot water, but—for the sake of future generations—you should always replant trees to replace any that you fell and take great care not to cut down and burn all the mature trees in an area.

KEEPING TRACK

If you are growing your own fruit and vegetables and keeping the odd chicken and a hive of bees more as a hobby than a business enterprise, there is no real need to keep detailed accounts.

If, however, you are running your operation on a larger scale, it is a good idea to follow some accounting routine, or else it is quite possible to delude yourself about how much money you are saving. You may also be able to get some tax benefits if you are selling any of your farm products.

Balancing the books

The cost of animal food, fertilizer, fuel, equipment, machinery, and the like must be balanced against the market value of the food you produce. By keeping detailed accounts, you will be able to pinpoint those areas that are really not economically viable, and either abandon them or alter them in some way.

It is important to remember that the first year or two often involve plenty of start-up costs and some inevitable failed projects, so do not expect a huge profit from the start. Finances are just one of the rewards of this kind of lifestyle and you will need to be prepared for a slow start, balanced out by fresh enthusiasm for your new working routine.

Quality and savings

You may well find that keeping livestock saves you very little money and may even cost you more than your old life of supermarket shopping. You will not be producing free food by any means, although your eggs, milk, and meat should cost you less than if you were buying them. What you have to remember is that you are producing food of infinitely better quality than it is possible to buy. The eggs are richer, the milk fresher, and the meat more tasty. Add to this the satisfaction you are getting from caring for the animals and watching them flourish.

Understanding the rhythm of life

It is a good idea to keep records of how many eggs you get each day, the numbers of young in a litter or brood, the date they were born, and so on. It might help, too, if you keep similar records of the fruit and vegetable production so that you can see what does particularly well, and notice when certain fruit trees or bushes start to become less productive and need replacing.

Records like this help you to understand your land and crops by noting patterns of planting times and places, the effects of weather in different years, successes, and failures that you can learn from. It is this kind of observation and practice that will ultimately make the difference between a cost-effective homestead and one that is little more than a money-waster.

STARTING OUT

Planning

Whatever the size of your property and the land you have available for home food production, careful planning and thoughtful use of the space can make a lot of difference to its productivity.

Even a comparatively small garden, well-planned and laid out, could prove more productive than a larger amount of land where things are just randomly placed. Think carefully at the outset about where you could put everything you want to include, even if you are not planning to do it all right away. Many features of the plan—the vegetable garden, orchard, pigsty, and goat shed, for example—are likely to be permanent sites, so be sure you have chosen the best possible place, both in terms of convenience to you and best conditions for them.

The following guidelines will help you to decide where to locate each productive element of your homestead, but remember that they are only that—guidelines. Your vegetable garden, for example, can be as large or as small as you want or are able to make it. In many areas, you can provide a family of four with fresh vegetables each week of the year from a garden that measures only 10 x 12 ft (3 x 4 m), space that can be found even in suburban yards by sacrificing some flower beds or part of the lawn.

If you have a larger property, you will need to decide what proportion of it to devote to each type of cultivation and to livestock. Think about the fruit and vegetables you and your family most enjoy eating and which ones will store best through the winter. Then allocate space accordingly to try to avoid a surplus of a crop that nobody likes or that will spoil before it can be eaten.

Making space for livestock

How much space you need for keeping chickens depends on whether they will have a permanent run or whether you have a portable coop (*see* page 166). Remember

Planning and organization are essential. Raised beds like these make it easy to measure out rows and rotate crops from year to year, while good paths give access for tending and harvesting your produce.

that you will need to go to the chicken coop regularly to feed and water the birds and collect their eggs, so consider how close and convenient you want them to the house against the disadvantages of noise from having them nearby.

Likewise with goats—are you going to take them out during the day and tether them on nearby land, if this is possible, or on an uncultivated area at the edge of your property? Or will you keep them permanently on one patch of ground? If kept in one place, they will need more space and supplementary feeding.

Start to plan on paper
Make a list of all the crops you want to grow and the livestock you want to keep. Note planting and harvesting times, how much space each crop requires, and any information about good companion plants that will help you to plan your plot. Then draw a scale plan of your land and sketch out areas for planting, storage, livestock, and any new structures. Don't forget to think about paths for easy access, making use of any existing features and hardscaping if you can before planning new construction.

Once you have a rough plan, pace it out on the ground itself. Try to imagine how each element will work together and how practical the space will be to use on a daily basis. Only when you are happy that you have it right should you start digging, fencing, and putting in permanent structures that will be difficult to relocate later on.

This careful preparation in the early stages means that you can be flexible about using the following planning advice to suit your own property and needs.

Think about everything you would like to grow and keep, including livestock like chickens, and whether they will be free-range or kept in an enclosure.

THE VEGETABLE GARDEN

The space you choose for growing vegetables should not be overhung or heavily shaded by large trees. Vegetables grow best if they are exposed to the maximum amount of sun possible; few will really flourish in shady conditions.

Shelter from strong winds, especially in coastal or open areas, will make your vegetable garden much easier to cultivate successfully. The best windbreaks are either fences, trees, or hedges. Leave space between any windbreak plants and the edges of the vegetable garden, so that you can trim any protruding roots from the windbreak each year. This will prevent them from competing with the edible plants for nutrients and moisture from the soil.

The size and shape of the garden will of course depend on the space you have available, as well as how much time you are willing to devote to working in it, but a rectangular shape is generally the easiest to subdivide. Access to your edibles will be easiest if the beds are surrounded by paths, ideally constructed from materials such as brick or paving slabs, rather than bare dirt, which may get muddy during winter or dusty and weedy in summer. Loose mulch such as bark chips can help keep down weeds on dirt or grass paths. Paths should be wide enough to take a wheelbarrow—typically 3 ft (1 m) or more wide.

Where you decide to put the vegetable garden in relation to the house is a matter of personal preference. Some people like to have it close by the kitchen door, others prefer a spot away from the house. In some cases, there may be only one suitable spot. However, keep in mind that it is probably better to have the vegetable garden closer to the house than the pigsty or the chicken run. You should also consider where you have ready access to water. If you have a greenhouse or cold frames, put them close to the vegetable garden for convenience. The sloping side of any cold frames should face south.

The compost pile should also be easy to reach, although some people prefer to keep it away from the house, as it can attract flies and may smell a little in hot weather. It does not have to be within the vegetable garden at all, but you will find it inconvenient if it is too far away.

Subdividing the garden

No matter what the overall shape of your property is, it's usually best to have an edible garden contained within an area shaped roughly as an overall rectangle or square. This is the most convenient shape to divide for a crop rotation plan (*see* page 40). Odd-shaped pieces of the property could be used for perennial crops, such as asparagus, berries, and artichokes.

There are advantages and disadvantages to planting rows running east/west or north/south, depending on your climate. Rows that run north/south get the maximum amount of sun, catching it on one side in the morning and on the other in the afternoon. Rows that run east/west tend to be shaded by those on either side for part of the day, and may be better in very hot climates. If your garden is on sloping ground, it is better to plant rows of crops or build raised beds across the slope, than up and down it.

Fruit trees can be grouped in an orchard or dotted around wherever you have room. Don't try to grow crops beneath them, as the shade will be too deep.

THE FRUIT GARDEN

Different considerations apply to the fruit garden and to the two types of fruit—soft and tree fruit. Although it is possible to dot berry bushes around the garden wherever there is space, you will find it much more convenient to keep them together.

When working in the garden or harvesting your fruit, your time will be used more efficiently if all the bushes are close together; it is also far quicker and easier to provide protection from birds or deer over a large clump of bushes, than to cover each plant individually, and this protection is a necessity in many gardens. If you are growing a lot of berries, such as raspberries or blackberries, consider growing them in a fruit cage, which provides full protection and easy access.

If the area you choose for fruit bushes is close to the vegetable garden, try to place it so that it will not cast shade over the growing vegetables. Some soft fruits—such as blackberries, loganberries, and raspberries—are somewhat shade-tolerant, so they can occupy shadier parts of the yard, leaving the sunny areas for plants that need full sun.

Remember that soft fruit bushes last for several years, so don't plant any in a place that you might want to change in a year or two's time.

Orchards of fruit trees

Tree fruits should not be planted close to the soft fruits, as they will soon overshadow them and rob them of important soil nutrients. Instead, incorporate fruit trees into your overall garden plan or allocate a large area to be a dedicated orchard. Fruit trees can be placed either where there is room for them or in a spot where you want to introduce some shade. They can be planted too, to improve the view from the house, perhaps by hiding or screening some unattractive feature.

Remember to choose tree fruits carefully, making sure they are on a suitable rootstock for your climate and for the height and spread of the tree you want (*see* page 113); otherwise you could end up with a giant that soon outgrows its space.

In cooler areas, heat-loving fruits such as apricots, peaches, grapes, and citrus can be placed against a south-facing wall. Walls also make good growing surfaces in small urban gardens, and can be used for growing espaliers of apples, cherries, and pears, or for some of the soft fruits that need training and supporting as they grow.

THE HERB GARDEN

Herbs might not seem essential to becoming self-sufficient, but growing herbs will not only allow you to add flavor to your food, but introduces delightful scents to your garden and will attract plenty of beneficial insects to help to keep plant pests under control without the use of chemical insecticides.

Because herbs are an additional bonus rather than a necessary part of your plan, the herb garden usually takes up the smallest amount of space. Herbs can be grown anywhere that is convenient, but are best sited close to the kitchen, so that they are easy to access when you are preparing dinner. Grow them in a small bed, so you can reach them all easily, and one with lots of sun—few herbs will do well in a shady spot with heavy, damp soil. If space is at a premium, you can tuck herbs into your flowerbeds or in containers, hanging baskets, or window boxes.

The herb bed can be as ornamental as it is useful. Here, the feathery fronds of a row of dill plants and their bright umbrella flowers are as pretty as any flower border—and edible, too.

DRAWING UP A PLAN

Before you begin to plant anything on your land, you must decide how it can be used to its best advantage.

A small plot for the bare essentials

If you don't have much room (below), you can devote your land almost entirely to growing vegetables and fruit. You may have room for some chickens to supply fresh eggs and meat, but probably not enough room to let the birds live free-range.

More space on a bigger property

A larger piece of land (opposite) allows for more features, such as a permanent pigsty and a movable chicken coop, so that the chickens can scratch around. The extra land means that you can grow a greater variety and get larger harvests of fruits and vegetables.

A hobby farm on a large property

This large plot (bottom) not only has space for pigs and chickens, but for goats, bees, or even a horse. The extra land allows for a system of crop rotation where there will always be one plot lying fallow, which can be used for grazing livestock.

A small plot, no bigger than a large backyard

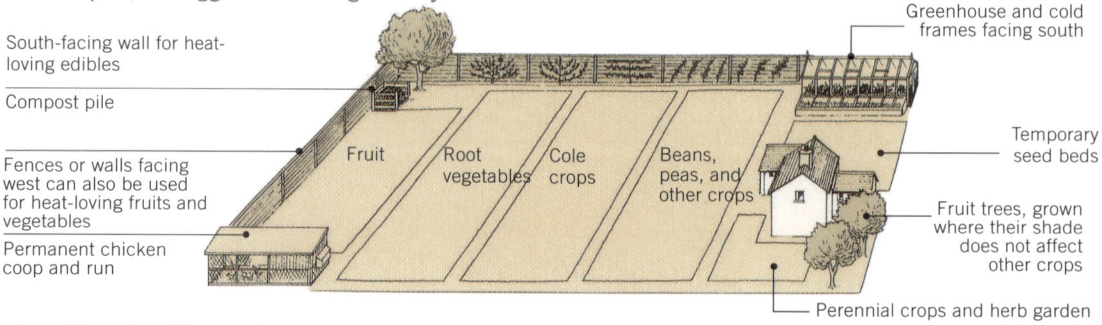

A large property has space for more variety

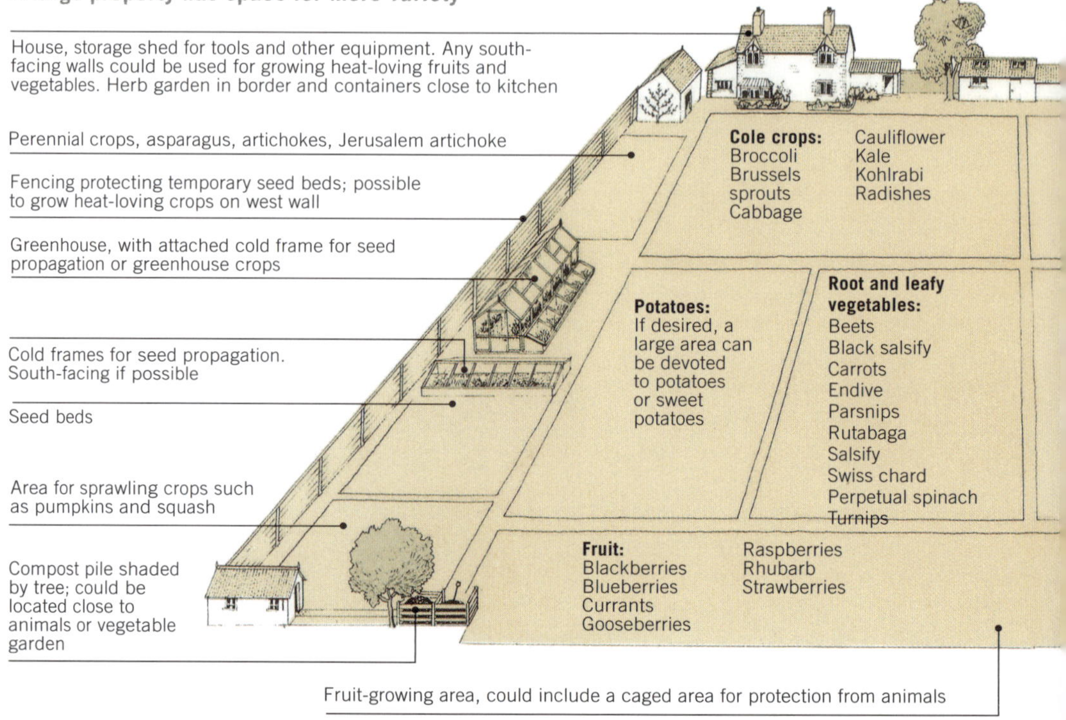

With more space, grow more of everything and add an orchard

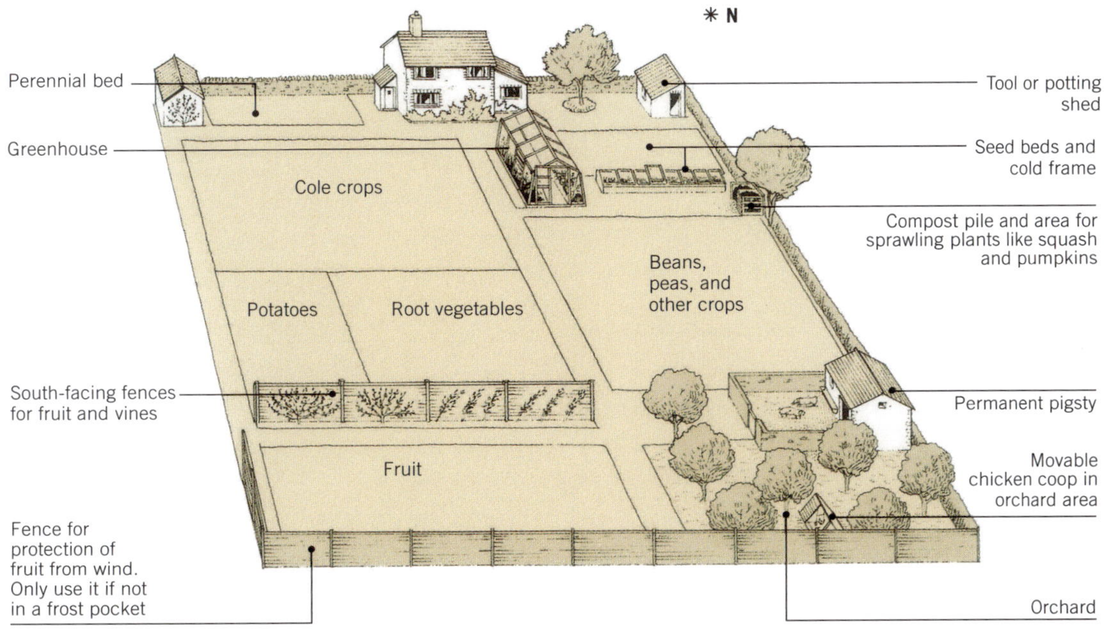

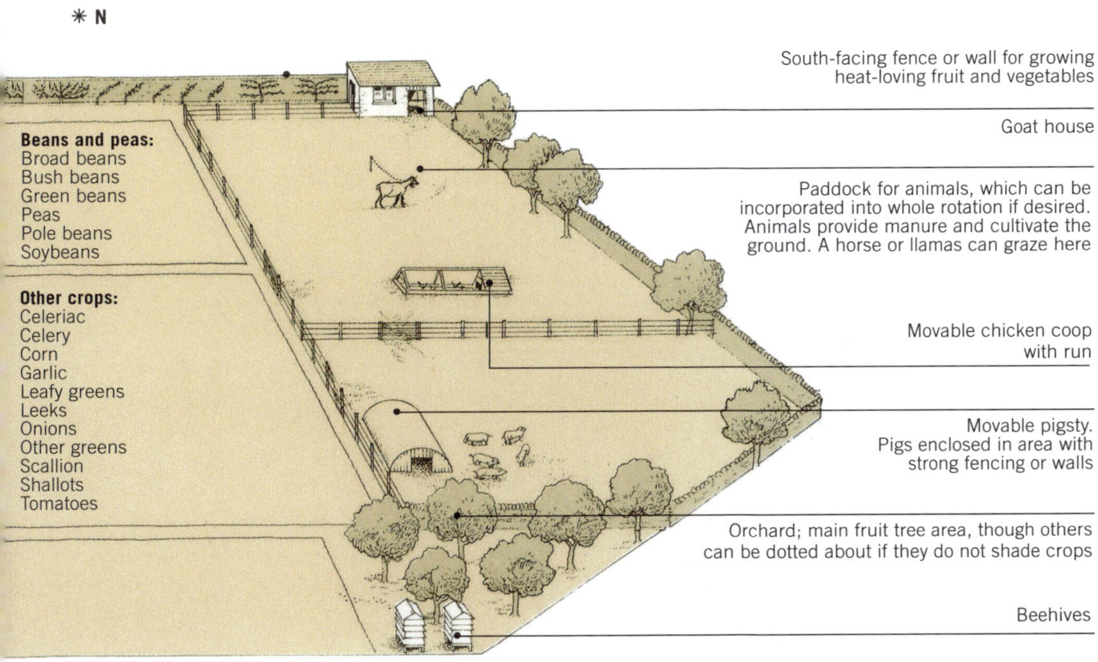

CHICKENS

Where you keep your chickens in the overall plan will depend as much as anything on which method you choose to keep them. Free-range chickens need only a permanent house and this can be sited anywhere. If there is a building suitable for a chicken coop already installed on the property when you take it over, then this is undoubtedly the best place to choose.

If you intend to keep your chickens in a permanent run (as opposed to free-range), it might be advisable to put this a little way from the house as it is not likely to be the most attractive of all your garden features.

The size of it is really up to you; allocate as much space as you feel you can spare. Common sense will tell you whether the chickens have enough room. In fact, they do not need a great deal of space in the run, although they will peck over as much land as you give them; it is far more important to ensure that you provide 4–6 in (10–15 cm) per bird around the feeding trough or hopper, and about 8 in (20 cm) per bird roosting space in the chicken coop (*see* page 166).

If you can allocate two patches of ground to the chickens, there is more chance of keeping both of them in better condition than if you have only one, as they can be rested in turn. Think of placing these two areas side by side; if there are two entrances to the coop, this can be used for both runs and the fence dividing them would provide a boundary for both patches.

OTHER POULTRY

Although chickens are the most common choice for keeping poultry, ducks, geese, and even turkeys are also popular and productive. The considerations for keeping them and allowing them sufficient space follow the same practical considerations as for chickens.

A gaggle of geese looks appealing, but be warned that geese are truly noisy birds, so take care where you keep them.

As all poultry must be kept indoors at night and let out by day, it may be more convenient to site them close to one another. It could make feeding a quicker operation, too.

If you want your geese to act as watchdogs, and make the loudest of all possible noises whenever anyone enters your property, put them near the driveway; if you want more peace and quiet, put them further away.

Ducks, of course, need a pond, though it doesn't have to be large. They need accommodation, too, but very little additional room, and are usually happy to share space with your chickens.

Never keep chickens and turkeys together as this can spread blackhead disease, which is fatal.

GOATS

A goat or two can make a worthwhile addition to your land. It may not be practical to keep a cow, but a goat is a far easier proposition and will provide you with milk and highly effective weed control. They are appealing creatures, too, and are not hard to accommodate, even in a relatively small yard.

If you intend to run the goat shed on a deep litter system—that is, to keep adding more and more bedding to the floor and then cleaning it out when it gets too deep (*see* page 162)—it is probably best to place the goats some way from the house where any smell it creates will be less of a problem.

There is no essential reason for the goats' living quarters to be near to the dairy, if this is a separate unit, although you will need to milk the goats twice daily, so the closer the two are, the more convenient the job will be—especially in bad weather.

Regulations and Licenses

To keep pigs, goats, or sheep—even just one animal—you must comply with any local regulations in your municipality. There may also be legal restrictions on moving livestock. (Most people can keep chickens and other poultry without restrictions.)

When you buy an animal, keep all receipts and any papers, such as vaccination records, that come from the breeder or seller, and continue to keep records throughout your ownership of the animal.

To find out more about rules and regulations for keeping livestock, contact your local Cooperative Extension Office or state, provincial, or municipal authorities.

There are no particular size requirements for a goat paddock, even if this is to be their permanent home, although it must be very strongly fenced if they are to be turned loose within it because goats are determined escapees. But goats are infinitely adaptable and will live happily in most conditions.

Making enough space

Having given them room to graze without bumping into one another all the time, the only consideration regarding the size of your paddock is how much you want to supplement their feeding. The smaller the grazing area, the more additional feed you will need to supply. Also, if the area is to remain as grass and not be stripped bare, it will need to be left fallow occasionally—just as with chicken runs. This means you must either allocate another patch of land, similarly fenced, to the goats, or you must have access to vacant lots or open ground where you can take them and tether them during the day.

If you tether goats, they should be moved twice a day as they will not eat plants where they have defecated, and you should not keep them too close to bushes or long patches of tough grass, where their rope could become entangled. If this happens there is a risk that the goats could strangle themselves.

The most important thing when preparing a paddock for keeping goats is to make sure that the fences are really strong.

In winter, it is best to keep goats on a piece of land that has been paved (again it does not have to be very big) if the ground is liable to become very muddy, as this could lead to various feet disorders.

PIGS

It is a myth that pigs are dirty, smelly creatures, but nonetheless, their food (which is usually root vegetables and fruit) is liable to smell. In addition, their run needs to include a muddy puddle and they are definitely noisy, so they are best kept at a distance from the house.

If you are keeping pigs to help you cultivate a very rough area of land, you will move them and their movable sty around as necessary. They will soon clear the land of scrubby undergrowth and even perennial weeds. It is more likely, though, that you will want to keep them in a sty with a permanent yard.

Pigs do not need much room if they are not relying on the land for their food. All they need is space to root around and to make a nice, deep, dust or mud hole.

This doesn't apply if you are planning to let your sow have a litter. She is likely to have about ten piglets and these have to be weaned when they are between five and eight weeks old. At this point, they will need separate accommodation from her, and depending on how long you mean to keep the young pigs, it may be necessary to split them up still further as they grow. Ten healthy, squealing piglets is a large number to keep in one fairly small yard, and overcrowding can lead to health problems and fighting.

Pigs are happy without a lot of space as long as they have a muddy spot to wallow in. Always offer them shelter from the sun.

BEES

These are the livestock that take up the least amount of space of all. Their only housing requirement from you is a hive.

There are some points to keep in mind when you are considering where to position your beehives, and these are discussed on page 176. Most people choose to put their hives at some distance from the house, so that well-trodden paths from the house to various parts of the property do not cross the bees' most usual flight path to and from the hives. It's often a good idea to have a screen of plants between the hives and busy parts of the property.

Think about your neighbors' property, too, to try to avoid having your bees flying regularly through your neighbor's patio or deck, where they might not be welcome.

Position beehives where the flight paths of the bees will not pass through your own most frequently used parts of the garden.

Clearing Overgrown Land

If the land you want to cultivate is a tangle of bushy undergrowth, you must clear it first. Pigs and goats are renowned as excellent ground-clearers, but in most cases you should probably clear the land with the help of some power equipment. The sweat and toil of clearing the land is hard but satisfying work.

The basic and most useful tools and equipment for clearing and turning over your land are discussed below. If you intend to keep goats, it may be worth getting them now, because they will do most of the work for you, eating their way through the undergrowth, even stripping the bark off trees and killing them, if you let them. Pigs will also clear land, but it will take either type of animal a while to graze through heavy overgrowth.

Clearing and felling

Seriously overgrown land may call for a backhoe or tractor, especially if you have large boulders or trees to remove. For tackling overgrown land by hand, you will need a fork, spade, scythe, pickaxe, shovel, axe, saw, and crowbar, as well as power tools such as a chainsaw, heavy-duty string trimmer, and maybe a brush machine and rototiller. Always wear boots to protect your feet and sturdy gloves, as well as eye protection.

Attack perennial weeds such as blackberry, thin-stemmed bushy clumps, and young saplings first. Remove all the fallen branches and other pieces of wood, chop them into manageable sizes, and store them for use as firewood.

Having cleared the ground of the undergrowth, decide if you want to remove any boulders and trees. These may be jobs for an expert, although instructions for felling a tree are given on page 203. Remember that you must also remove the stump from the ground and doing this is a major job in itself. Stumps can be tractored out with heavy-duty chains or ground out of the earth with a large drill-like tool (again, probably jobs for an expert). If you opt for grinding the stump, make sure that you ask the arborist to grind it deeply enough. For cultivated ground you will need to go at least 12 in (30 cm) below the surface.

If you decide to remove a stump by hand, the best approach is to dig it out with a mattock and a spade, but this is a long job. Alternatively, you can treat stumps

If you have a large area to clear and prepare for growing, a rototiller makes fast work of the job. Try to hire or borrow one, since you will not use it often.

Use a pickax and shovel to dig out roots and stumps of shrubs and to break up compacted dirt or hardpan. Hard work now means better crops later.

with a chemical to speed up decomposition. Use a commercial stump killer, and make sure to wear protective clothing and follow the manufacturer's instructions carefully.

MOVING A BOULDER

It may be possible to work the position of a large, deeply set boulder into your plan, but if it is in the way of a proposed path or building, or if it lies in one of your planned planting areas, it will need to be moved. Dig around it to establish its full size and see how best to tackle it. You may be able to pull a boulder out of a shallow dip with a rope, but well-buried stones must be levered out.

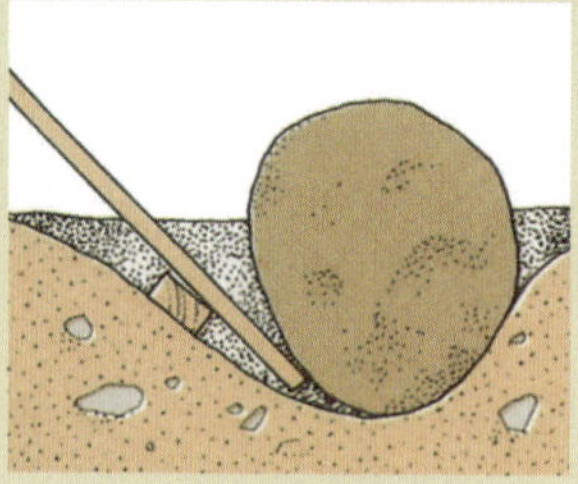

1 Using a strong pole, work it under the boulder. Put a block or brick under the pole, and use it for leverage.

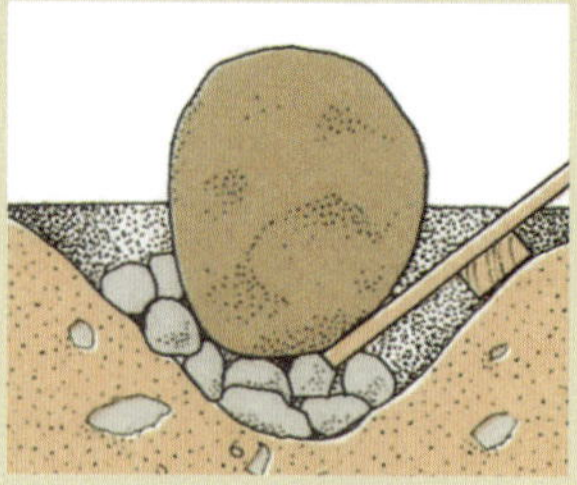

2 Still holding the pole, begin to place large stones under the boulder to raise it from the bottom of the hole.

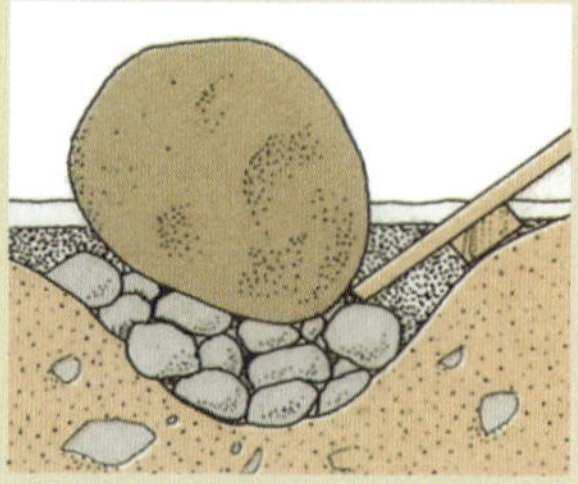

3 Having raised the boulder, lever it in the same way from the other side. Continue adding stones to raise it.

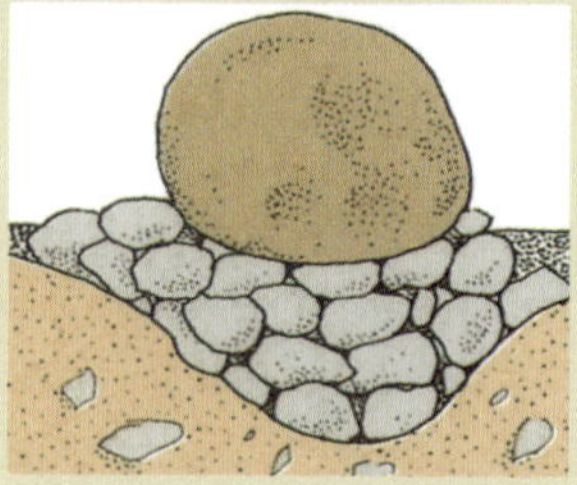

4 Finally the stones will bring the boulder up to the level of the ground, at which point you can roll it away.

Breaking up the soil

Neglected land is best broken up for initial cultivation by a rototiller. Before you can do this, all large stones, boulders, construction debris, bits of brick and wire, or other garbage must be removed, or they will damage the tiller blades. Large boulders may have to be levered up out of the ground (above right) and this is a two-person job. Remember to lift all heavy items carefully. Bend your knees and lift with your back straight, rather than bending over the object and straining your back muscles. Very heavy boulders could be hauled away by towing them with a tractor or a truck with heavy-duty chains. Do not necessarily take all the debris off to the dump, but think whether you can reuse it. Bricks and large stones or stumps can be used to line paths, build raised beds, and many other jobs around the property. Keep anything that might be useful in an inconspicuous place.

Once the ground is cleared, rototill the areas you want to cultivate or dig them over thoroughly by hand (*see* page 28). Again, goats allowed to graze the area for a week or so will remove many harmful weeds—at least disposing of seeds, if not the deep-seated roots of perennial plants. If the weed growth is very heavy, you can work over the entire area a few times with a weed flamer—but only if you know how to use one properly, and the weeds are really persistent. This will destroy weed seeds, but also the natural organic matter that will help to improve the soil.

Both gas-powered and electric rototillers can be rented, but their basic purpose is the same—to turn over the soil. Do not rush the machine over the surface, as it will not dig deeply enough.

Improving the Soil

Soil is always made up of a series of layers. The top layer, or topsoil, is the one that most affects the successful growth of all edible plants.

Topsoil can be categorized into three main groups, although these could each be endlessly categorized, as few soils fall neatly into one simple category. In general, most soils are sandy, loam, or clay. One method of assessing your soil type is to mix a handful with water, shake it well, and leave it in a glass jar overnight (right).

Your local Cooperative Extension Office is another resource to help you identify the soils in your area and the plants that are best suited to the site.

THE GLASS JAR TEST FOR SOIL TYPE

Before drawing up your planting plan, you must discover what sort of topsoil you have, because some vegetables and fruits will only grow well in particular types of soil. There is no point in wasting time and money in growing produce in soil to which it is not suited.

Shake a small quantity of your garden soil in a jam jar of water and leave it to settle overnight. Gravel and coarse sand will sink to the bottom and will be topped by a layer of gritty material. Above this will be the clay. If the gravel and sand makes the largest layer, the soil is sandy. The middle layer indicates the amount of loam present; if it makes up about 40 percent in all, the soil is a good loam. If the top layer of clay is equal in depth to the other two layers together, you have a clay soil.

Sandy soil

This is light and easy to dig and warms up quickly in spring. Sandy soil offers little resistance to a shovel and is not heavy or sticky to lift. Pick up a handful; it should feel gritty and slip easily through your fingers. In windy areas, the surface of sandy soil is subject to erosion, and water drains through it so quickly that it can become very dry. In addition, vital nutrients are washed away. The lost nutrients should be replaced and the ground made more water-retentive by digging in large quantities of organic matter.

Loam

This is the ideal soil for all types of cultivation. It is darker than sandy soil (owing to its humus content) and crumbly rather than gritty. A handful rubbed between the fingers will feel smooth; moisten it slightly and it should not feel gritty, nor should it stick in lumps. Loam retains moisture and nutrients well.

Clay

Clay soils are heavy to work, sticking to tools and your boots. A handful pressed between your fingers will stick together in a lump. It drains slowly and after a heavy rainfall, puddles will remain on the surface of the ground for a long time. Clay soils retain nutrients well but they are slow to warm up in spring.

In a long, dry spell, clay soils can become very hard. You can improve the texture of clay soils by digging thoroughly in fall, leaving large lumps on the surface that will break up in the winter frosts. Dig in large quantities of any available organic matter—such as well-rotted compost, farmyard manure, or leaf mold—to break down over winter. When the soil has dried out in the spring, it can be forked down and raked to a fairly fine tilth.

Hardpan

A completely impenetrable layer of soil is found naturally in some regions of the West and Southwest (where it is called *caliche*). It can also be found in new residential developments where the topsoil has been stripped away or the ground has been compacted by heavy equipment. Hardpan may call for additional drainage (see page 27), or you may need a professional to break up and amend the soil. In worst cases, plants may best be grown in raised beds, just as you may need to do if your are in an area with shallow soils over bedrock.

TESTING THE SOIL PH

In addition to their type, soils are either alkaline, neutral, or acidic and this will also affect which crops will grow well. You can assess your soil's acidity by using a soil-testing kit, available at any garden center.

The soil tested will register a number on the pH scale. A neutral soil has a pH of 7; above this the soil is alkaline and below it, it is acidic. Alkaline soils are most often found in arid parts of the continent, where rainfall is light. Acidic soils are more common in areas with high rainfall, such as the Pacific Northwest. These soils also tend to be low in fertility.

Most vegetables grow best in a slightly acid soil with a pH reading of 6.5. Cole crops prefer a pH of 7 to 7.5, while squash and tomatoes like a pH of 5 to 5.5. Most fruit will be most likely to thrive in a neutral or slightly acidic soil.

Alkaline soils may be amended over time with sulfur, or with ongoing amendment with organic matter.

Overly acidic soils can be corrected by adding quantities of calcium carbonate (lime). It should be sprinkled over the soil as long as possible before growing crops, but after the ground has been dug in the fall or winter and left to be washed through the ground by winter rains.

A good soil is rich and crumbly, with plenty of organic matter. It should feel smooth, not gritty.

USING A PH AND NUTRIENT TESTING KIT

As well as discovering the type of soil you have, you can find its pH and nutrient levels using a soil-testing kit. Take small amounts of soil from different parts of your property and put them in test tubes. By adding different solutions, you can determine the pH and the level of the three main nutrients required for healthy plant growth: nitrogen, phosphorus, and potassium. A more detailed soil analysis may be provided by a commercial soil-testing company, or through your local Cooperative Extension Office.

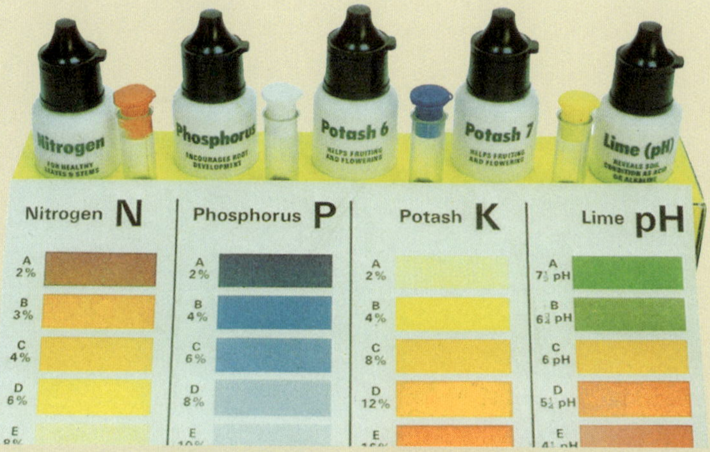

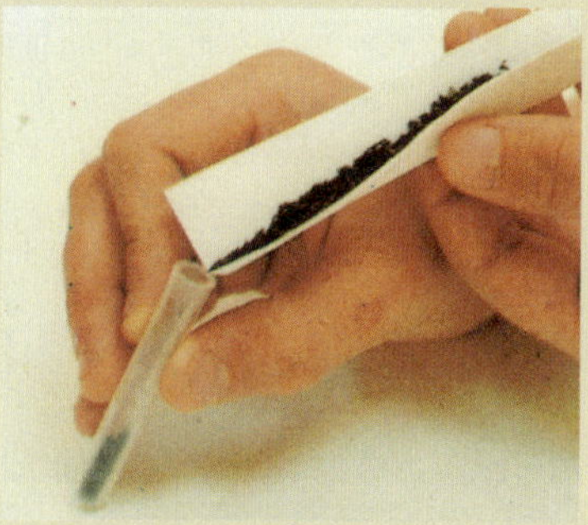

1 Take samples of soil from different areas of your property, allow them to dry, remove any debris, and crumble the soils. Quarter-fill each test tube with different soils, then label the tube.

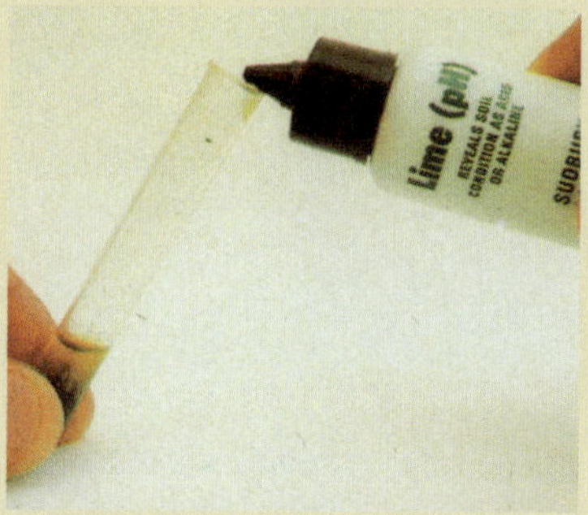

2 Holding each test tube carefully, add a few drops of solution, according to the kit instructions. Shake the mixture for 30 seconds and then leave for 10 minutes.

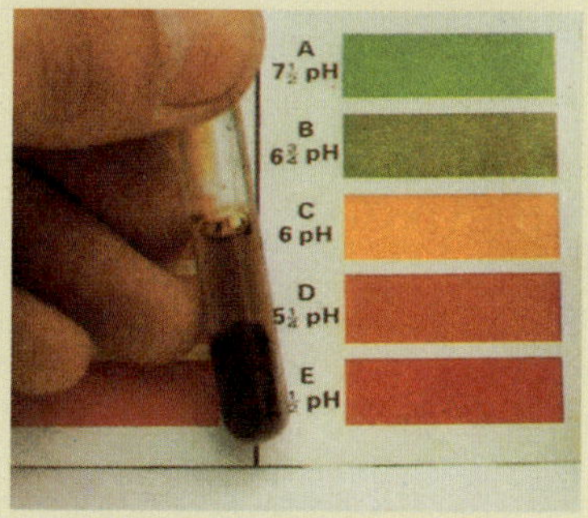

3 When the soil has settled to the bottom of the tube, match the color of the solution to the appropriate chart. The result gives the pH level of the soil or the percentage of each chemical nutrient.

Composting and Amending

As they grow, all plants take nutrients from the soil. Unless these are replaced, plants will grow poorly. Whenever possible, soil should be revitalized and its stocks of plant food replenished using organic matter, soil amendments, and homemade compost.

Inorganic or chemical fertilizers generally produce more immediate results than compost or organic amendments, but in some cases they can cause lasting damage to the soil. It is usually possible to keep the soil healthy by adding regular applications of well-rotted manure and compost, usually in fall. Mulches of these materials—a layer spread over the top of the soil—also help to prevent the soil from drying out in hot or windy weather, and they can help to control weeds.

HOMEMADE COMPOST

This is produced by combining together all kinds of organic waste—dead plants and deadheaded rotten flowers, kitchen scraps, fruit skins, crushed eggshells, hedge and grass clippings, and fallen leaves.

Compost ingredients should be layered in a suitable container interspersed occasionally with an activator, which helps to speed the decomposing process by feeding the bacteria. The activator that helps with decomposition is nitrogen, which can be supplied organically with a layer of poultry or other animal manure, fish meal, dried blood, or grass clippings. You can also buy compost activators at garden centers and add these to your pile to speed up the composting process.

What to add to the pile

Begin the compost pile with a layer of coarse twigs or stems to allow the air to circulate from the bottom (air is essential for the decomposing process), then add organic waste in even layers, alternating wet waste with dry waste like leaves. Press the layers down and dampen them if they are very dry.

Do not compost perennial weeds or annuals that are full of seeds, any diseased materials (such as the infected roots of a cabbage, for example), or scraps of food that are greasy—the seeds will grow and the infections spread. Anything that is very thick or woody should be chopped first. Lawn clippings should

COMPOST BINS

Making your own compost is an easy, but vital part of vegetable gardening. The compost should be contained in some form of bin, whether it is commercial or homemade. If making one yourself, remember to leave holes in the sides to allow air to circulate. While the compost is rotting, cover it with plastic sheeting or landscape fabric to preserve warmth and moisture. Turn the compost at least once a month. Try to have at least two piles at one time—one rotting and one in use.

When one bin is full, leave it to rot and use it while you fill the next

Some bins allow you to remove compost from the base when it becomes ready to use

A simple wire box is quick to construct and suitable for making compost or leaf mold

be mixed with other materials rather than spread in a solid layer as this can form an effective air-excluding barrier, and prevent the bacteria from working.

Water a compost pile in hot, dry weather to keep it moist and cover it with a plastic sheet or landscape fabric to insulate it and help it to retain moisture.

Waiting for it to "cook"

A compost pile is most successful if built quickly, preferably in the spring and summer when the warmer atmosphere speeds up the decomposing process. At this time of year it will be ready in a month or two, at which point it will be brown, moist, and crumbly, with a uniform appearance and texture throughout. It will take at least twice as long to rot down in cool weather.

You can make compost in a loose pile in an out-of-the-way corner, in homemade bins (*see* page 25) or commercial bins or cones. Plastic composters are often the quickest method—they are usually black, to retain heat effectively—but they do not hold large quantities and you are likely to need several of them.

An open compost pile in a wooden frame will hold much more than a plastic bin, but take longer to decompose.

Because they work fast, you do not need to turn the contents as they rot.

Worm composters, or wormeries, are useful for breaking down kitchen scraps before adding them to the larger compost pile. Wormeries require red wiggler worms, available from specialty suppliers, rather than earthworms.

FARMYARD MANURE

This is most likely to be horse, cow, pig, or poultry manure, mixed with the original bedding of straw or wood shavings, and may be used to add valuable humus to the soil.

Manure is very valuable to the fruit and vegetable gardener, although it is seldom as rich as well-made compost. This is especially true if the manure has been stacked out in the open, because the rain may have washed away many of the soluble nutrients. Poultry manure is usually best added to a compost pile, rather than straight to the soil.

OTHER ORGANIC AMENDMENTS

You can also enrich your soil by digging in one of the natural additives listed below, either made in your garden or commercial.

Leafmold

This is produced by composting dead leaves and adds valuable humus to the soil. If you do not have enough leaves to warrant composting them on their own, add them to the general compost pile. If you have a lot of leaves, layer them in a pile, press them down well, and dampen them if they are very dry. Do not make a pile more than 3 ft (1 m) high and turn it three times a year. It will take around eighteen months to decompose, by which time it will have shrunk to half its size.

Seaweed

If you live in a coastal area, you could collect seaweed, which is an excellent source of plant nutrients. Leave it outside for one or two rainfalls to wash away the salt then dig into the soil in fall or winter.

Peat

Dark brown or black-colored peats are useful for enriching the humus content of the soil and, therefore its drainage, but contain little in the way of plant food. Lighter-colored peat is not as well decomposed and will be less rich in nutrients.

Bonemeal

Dried blood, fish meal, and bonemeal are all used as a top-dressing, or added at the bottom of a planting hole to enrich the soil at planting time. They are valuable, but expensive and add no bulk or humus to the soil, but enrich its content.

Improving Drainage

Soil with poor drainage can often be improved by regular digging and by incorporating lots of organic matter.

If you have improved the humus content of your soil and the ground still appears to be waterlogged, or if water rests for days on the surface, try raising the soil level by still more digging and adding bulky organic matter.

Better drainage can sometimes be achieved by draining the ground above the plot. Dig a ditch at the end of this area, sloping it down into a sump (a deep hole, the bottom of which should be beneath the level of the water table) filled with rubble. Mark the position of the sump in some way, as it will probably need to be periodically cleaned out to remove silt and other debris.

Draining with a sump

A trench drain can be dug in the subsoil, sloping from the highest to the lowest point and leading to a sump. It must be as deep as it is wide and have a layer of stones or rubble in the bottom, with smaller pieces of gravel on top. The sub- and topsoil are then replaced.

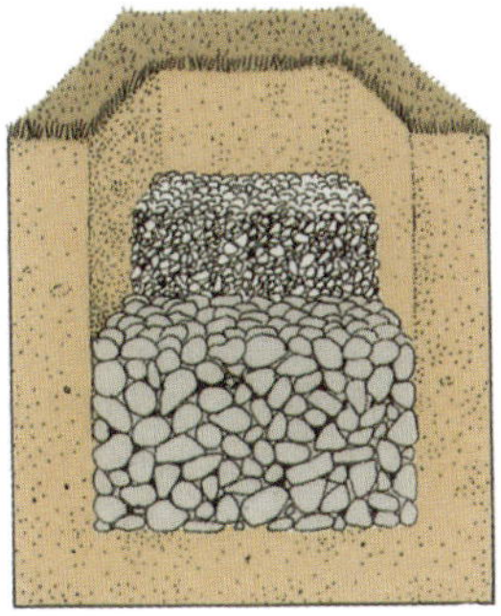

Draining with a pipe

To make a sump drainage system more effective, lay a pipe in the trench before filling it up with rubble and gravel. Use rigid PVC drainage pipe that has holes along its length. The water soaks through the holes and drains away to lower ground.

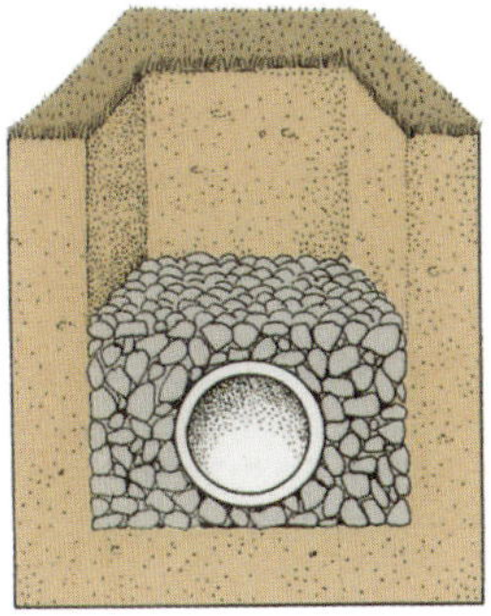

The herringbone system

A professionally installed trench drain system may be arranged in a herringbone pattern (left) of smaller drainage pipes leading into a main pipe and, from there, into a gravel-filled pit.

The herringbone drainage system

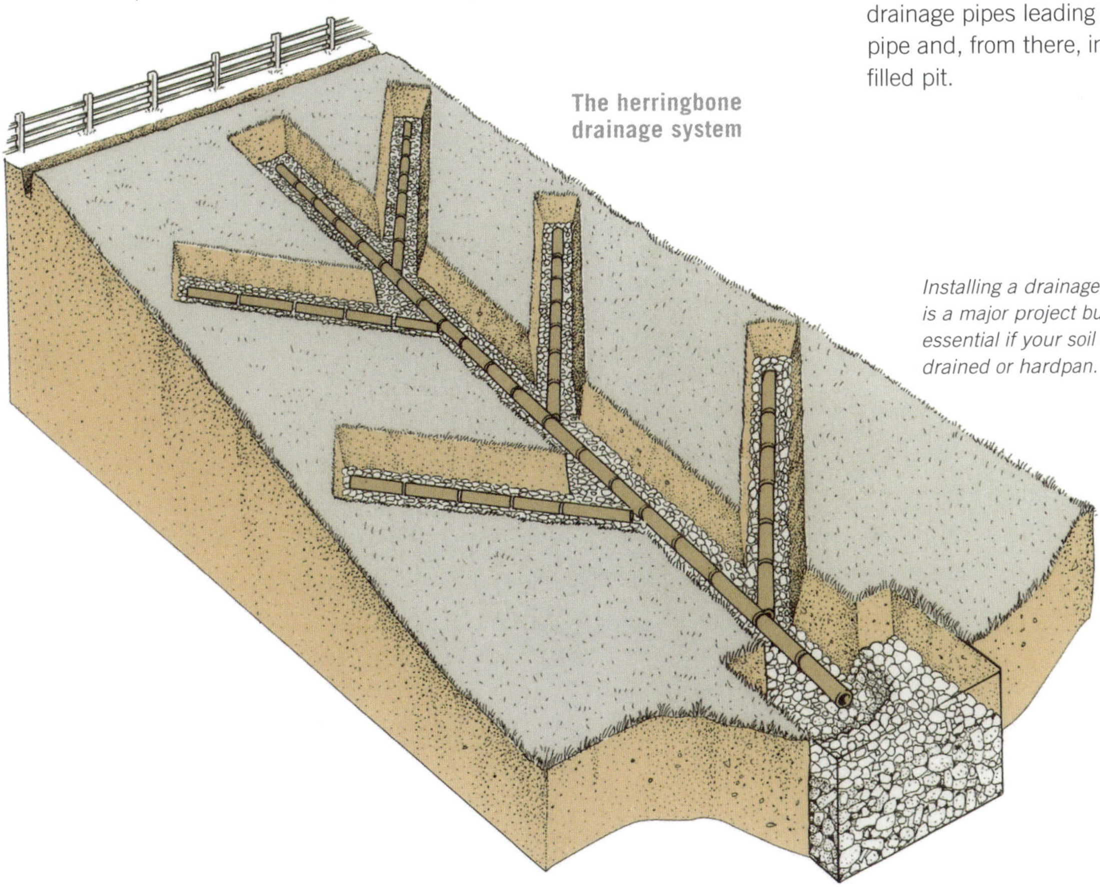

Installing a drainage system is a major project but may be essential if your soil is badly drained or hardpan.

Before you start digging a large area, choose a shovel that is the right height and has a comfortable handle. Wear thick-soled boots to cushion your feet.

Digging

Turning the soil is an essential part of your yearly cultivation. It improves the soil's condition by aerating it and making drainage more efficient, while also making it easier for moisture to reach and be absorbed by the plant roots. It also provides an opportunity to clear the soil of old plant roots and weeds, and to incorporate organic matter into the growing area to replenish lost nutrients.

The time to dig heavy clay soils is in the fall; not only will they be harder to dig later in the year, but they will not reap the benefit of the winter frosts in breaking up lumps of dirt. Don't knock back the soil as you dig, but leave it in large lumps on the surface to expose the maximum surface area to the coming frosts.

If you miss the opportunity to dig in fall, wait until the soil dries out in spring. You can dig lighter soils at any time the soil can be worked.

Single or double digging?

A traditional way to prepare the soil for growing fruit and vegetables is by single or double digging (*see* below and opposite). Both methods can be strenuous, particularly if you tackle the job incorrectly. Follow the method shown in the diagrams and always lift less dirt on the shovel than you think you can manage. Stand upright, stretching your spine at frequent intervals, and work slowly, especially if you are not used to regular digging.

Single digging is adequate for most purposes. Double digging is usually done on long-neglected ground. Heavy clay soils can benefit from double digging, as it will help to improve drainage. A new vegetable garden will yield better crops from the outset if it has been thoroughly double-dug.

As you dig, remove the roots of any perennial weeds. Annual weeds can be dug into the soil to add to the humus as they rot down.

Single digging

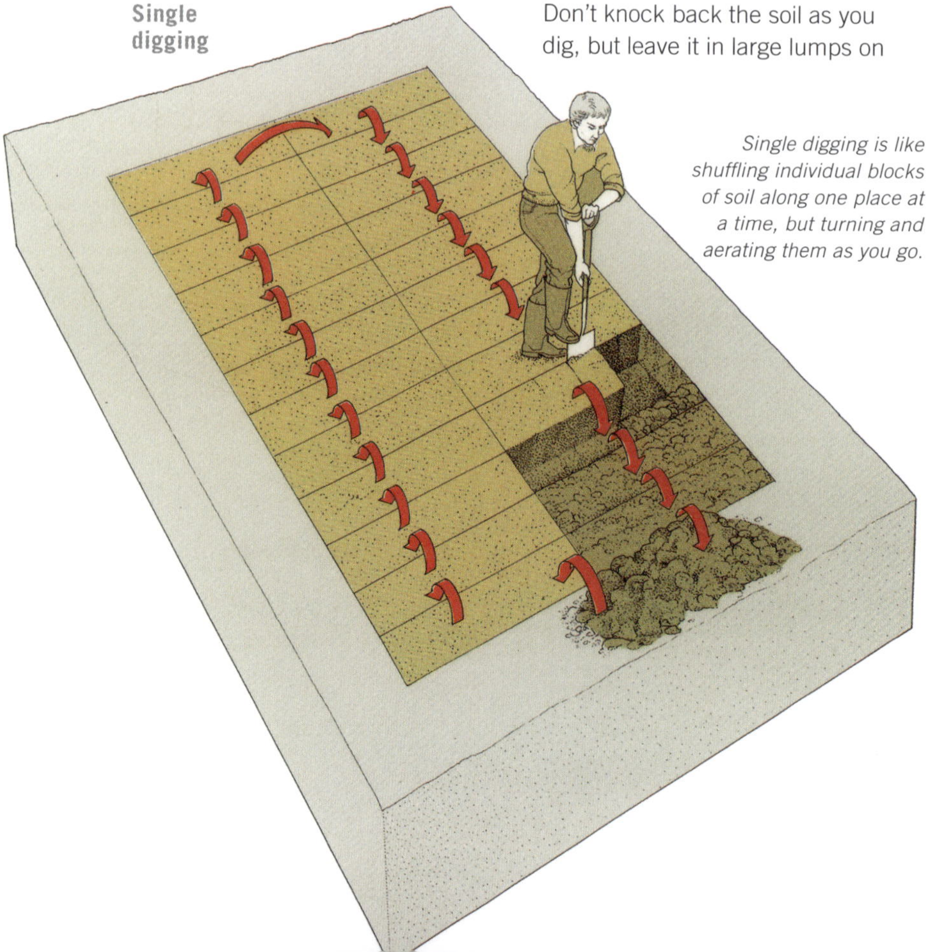

Single digging is like shuffling individual blocks of soil along one place at a time, but turning and aerating them as you go.

SINGLE DIGGING

Work methodically and do not attempt to cover too large an area at any one time, particularly if you are not used to digging.

Mark the area into two rectangular strips. Dig out a trench on one strip, the depth of the shovel and a little wider than its width. Put the soil in a wheelbarrow or just beyond the top of the next strip until you need it, at the very end of the process.

Dig narrow rows across the first strip, working backward to avoid stepping on newly dug ground. Mark a line across the width each time with the edge of the shovel, making each row 4–6 in (10–15 cm) wide—it is important to keep the trenches equal in size. Throw the soil forward into the previously dug trench, tilting the spade to the side to turn the soil as you go.

When you reach the end of the first strip, fill the remaining trench with soil taken from a trench at the same end of the adjoining strip. Work back down the other side of the plot in the same way, but the opposite direction, filling the final trench with the soil taken from the first trench you dug.

DOUBLE DIGGING

When you need to improve the basic structure or nutrient content of neglected soil, double digging is the best technique to use. It is hard work, but you will be rewarded in much better yields than if you had not done the job.

Use string to mark out the area into two rectangular strips, as you would for single digging. Dig the first strip the depth of the shovel, but twice the width. Remove the soil.

Dig into the soil in the bottom of the trench, to loosen it without removing it, then add an even layer of organic matter. Dig the next row, making it the depth and width of the shovel, throwing soil forwards to fill the first trench. Dig out two rows of topsoil to form a new trench, of two shovels' width, and continue to work your way along the bed, digging in organic matter at the bottom of each trench as you go.

DOUBLE DIGGING TO IMPROVE THE SOIL

Double digging incorporates organic matter into the soil to enrich ground that has been neglected or exhausted. It will also improve the texture and drainage of heavy soils. Work methodically, moving backward along the area.

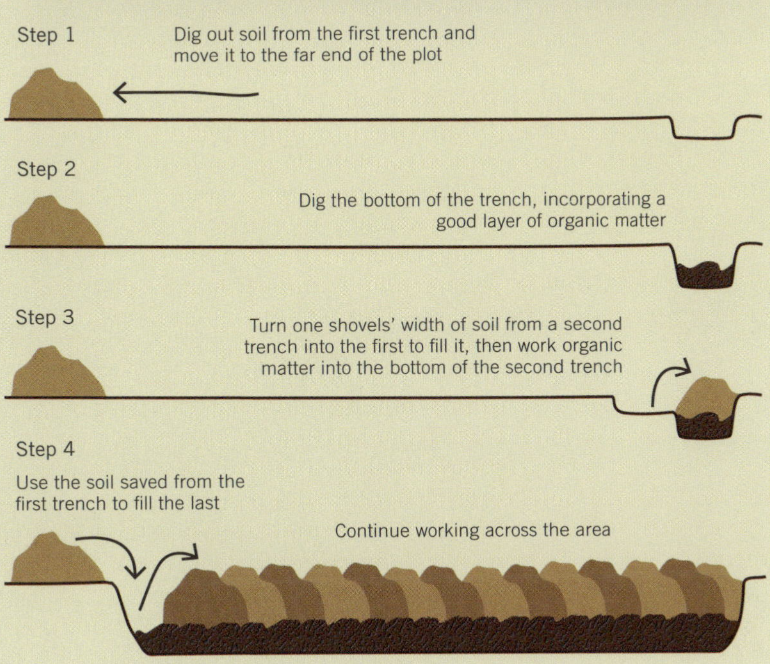

Fences and Other Boundaries

Even the smallest homestead will need fences—if only to mark its boundaries. If you're lucky you may already have sturdy walls or fences around your property, but if not, choose wisely and take care to erect them correctly and your boundaries will be trouble-free for many years.

Fences have all sorts of other uses beyond being simple boundary markers. They can act as wind-breaks to protect delicate fruit bushes or vegetables. They can form a screen to hide untidy areas of the garden, such as the compost pile, or to keep bees, perhaps, away from certain areas. They are essential for containing livestock such as chickens, goats, or pigs, and for keeping out other animals, such as deer. They can even act as supports for grapes, kiwis, or other climbers.

There are many different types of fencing. Some of the most common utilitarian options are illustrated on the next page, but there are others you can consider, including the solid lumber fence panels sold in standard sizes in most garden centers and home supply stores.

The cheapest of these panels may not stand up to high winds and will present little challenge to a determined goat with a view to escape, but they make adequate boundaries for areas that are not exposed to rough weather and they are both quick and easy to erect. Screw them to strong wooden posts

Post-and-rail fences are strong enough to contain boisterous livestock, but will not keep hungry deer away from your vegetables.

set in concrete footings that extend beneath the frost line in your area.

Think about form and function
Consider what function you want the fence to serve before selecting the type of fencing to use. This way you can choose the most suitable solution for the job. Of course, cost is a consideration, but do not necessarily choose the cheapest option. If it is not long-lasting, it will ultimately cost you more by having to replace it sooner than you would have done with a better initial fence.

Remember that boundary fences often have to keep pets and children in, as well as other people's livestock and destructive wild pests, such as deer, out. A post-and-rail fence makes an excellent barrier for containing livestock, but will do nothing to keep deer out of a vegetable patch.

Bear in mind, too, that solid fencing will block out all light from the area it casts into shadow, putting nearby fruit and vegetable plants into deep shade. Above all, any fence you erect is always only as good as its weakest point, so make sure that it is well built, with solid posts and secure footings.

Hedges and walls
A hedge is one of the most attractive of all boundaries and can form an effective barrier, too. Mountain laurel, holly, cedar, and yew make good thick evergreen hedges. But you must protect young hedges from animals such as goats, who will destroy them before they have a chance to establish themselves.

As a hedge grows, it can be "laid" to weave it into a more solid barrier by cutting into the base of growing stems and bending them over. In some cases, you can do this without any additional support; in others, separate upright stakes are driven into the ground and the living branches woven between them.

High walls are unbeatable for keeping pets, livestock, and children in, and unwanted intruders out, but they can present a rather stark appearance. Attaching a wooden trellis, or stringing strong wire between vine eyes hammered or screwed into the wall makes it easier to grow climbing vines to soften the appearance of the wall.

A south-facing wall is particularly useful for growing fruit such as peaches, grapes, kiwis, and figs.

ELECTRIC FENCING

An electric fence can be useful for keeping pigs and goats in a contained area and it is easy to move from place to place if you vary your animals' grazing area.

Fences used to contain livestock must always have some sort of gate incorporated into them to give you access to feed, clean, and inspect the animals. This is easy to incorporate into an electric fence. Drive two strong posts into the ground wide enough apart for your access, then cut the wire between them and slide a plastic handle onto one end. Twist the final piece of this wire into a hook and twist the wire on the adjoining post into a loop. When you hook these together, the circuit will be reconnected.

Pigs will need a two-stranded electric fence to contain them securely; goats will need three, but the bottom wire does not have to be live. Position the live wire at around the nose height of the animals it is containing. They will very quickly learn to stay away, rather than getting repeatedly shocked by the wire. Make sure to clear any long grass around the site of the wire, because if it touches the electric fence it will allow some of the electricity to "drain" into the ground.

An electric fence is an effective way of keeping livestock in their allocated space. Always put up a visible warning notice to protect passing visitors, especially children.

POST-AND-RAIL FENCING

There are many variations on the traditional style of post-and-rail fencing. They are one of the most attractive forms of fencing and form an extremely secure barrier, but once built, they are permanently positioned, so you cannot easily move them to adjust the use of the land they contain. They are also an expensive choice.

All post-and-rail fences are built by securing cross rails to sturdy vertical posts. You can use rough split rails or dimension lumber. The rails can be attached alternately to opposite sides of the vertical posts to give a different effect.

Post-and-wire structures make excellent supports for climbing fruit canes and vines. Use them together to make a productive and attractive boundary.

TYPES OF FENCES AND BOUNDARIES

The practicalities of fences and walls are important, but your boundaries should also blend in with your garden's surroundings. A dry stone wall might look out of place in an urban setting, and barbed wire is a poor choice for a suburban neighborhood.

Dry stone walls

No mortar is used in the construction. They are attractive and very stable if properly built. Semidry stone walls, in which mortar is used in the center, where it is hidden from view, are stronger, while retaining the dry stone effect.

Wattle hurdles

Can be purchased as panels or you can make them yourself by weaving split willow branches between vertical posts. They provide a more solid-looking fence than ordinary hurdles (below), but are used in the same circumstances.

Post-and-rail fence

Easy to build, although a neater appearance can be achieved by cutting the ends of the horizontal rails at a 45° angle and splicing these together to form a continuous rail. In some fencing, mortise joints are used to join the horizontals to the verticals.

Wire fences

Often used for containing livestock quickly and cheaply. Widely spaced hog wire is very strong; closer mesh chicken wire is easier to handle and perfectly adequate for poultry.

Hurdles

Very useful in making quick, temporary fencing to contain young livestock, for example. When not needed, hurdles can be neatly stacked against the wall of a shed or garage, where they will not take up much room.

Barbed wire fencing

There is rarely a good reason to choose barbed wire. It looks unattractive and can inflict injuries on livestock. In most circumstances, ordinary wire fencing will provide just as effective a barrier.

Tools and Equipment

The following selection of tools are those you are likely to find most useful in fruit and vegetable production. Start with a few core essentials and spend as much as you can afford, then add to your toolshed bit by bit or if you need a particular tool for a job.

When buying gardening tools, it is usually true that you get what you pay for. In other words, the more expensive ones will generally be made of better, stronger material, and will not only last longer, but will also do their intended job more efficiently. As with all garden tools, it is false economy to buy poorly made tools that will break much sooner than their more expensive, but better produced counterparts.

Do not make the mistake, though, of buying sophisticated equipment when much simpler tools will do the job equally well. Examine tools carefully before you buy, feeling to make sure the grip or handle is comfortable to hold and that they are the right weight for you.

Having bought good equipment, make sure you keep it in good condition. Scrape off all clinging soil before putting the tools away, and store them neatly, preferably by suspending them from hooks on the shed or garage wall. Keep metal parts oiled and greased, and sharpen all blades once a year.

Basic toolkit

These tools will form the basis of your arsenal of equipment. Use this as your "start-up shopping list" and add to your collection as your experience grows.

- Spade and shovel
- Garden fork
- Rake
- Cultivators
- Dutch hoe
- Draw hoe
- Hand trowel and fork
- Broom
- Containers
- Garden string and stakes
- Dibber
- Sprayer
- Watering can
- Hose; hose-end sprayer
- Hand pruners
- Loppers
- Ripsaw
- Pruning saw
- Pickaxe
- Wheelbarrow
- Buckets

The Greenhouse

A greenhouse is by no means an essential item if you want to grow your own garden produce but in colder regions it can lengthen the growing season and widen the choice of crops that you can cultivate.

If you don't already have a greenhouse, the cost of buying and heating one (if you choose to do so) should be carefully weighed against the value of the crops you will produce. A greenhouse has two main functions: it enables you to produce a variety of crops (tomatoes and peppers, for example) outside the usual growing season, or to grow specialty crops that struggle outdoors in cooler regions.

There are three main types of greenhouse and your choice may well be guided by the site you have available for it, but conventional ridge-roofed greenhouses are still considered by many to be the best shape of all. Buy the largest you can afford or accommodate—you are unlikely to wish you had less space.

Span or gable roof

This is the most popular and useful greenhouse choice for vegetable production. The walls can be glazed from top to bottom, or may have knee walls of 3 ft (1 m) or so—typically concrete block, stone, solid wood, or aluminum, depending on the rest of the greenhouse construction.

Lean-to

A simpler structure that is built against a house or garage wall. The wall absorbs heat during the day and releases it into the greenhouse at night.

A lean-to can also be built beside a potting shed, to keep tools and equipment close by. A disadvantage of a lean-to design is that it has light coming from only one side, so it is best constructed against a south- or west-facing wall. A lean-to is ideal for starting seeds, or growing tender crops like tomatoes.

Dutch gable roof

The sloping sides are constructed of large panes of glazing and attract the maximum amount of light.

Variations on the basic designs

Even among these three basic shapes there are many different designs, and it is also possible to get circular or hexagonal-shaped greenhouses. These look attractive and are extremely practical in many ways because they allow in as much light as possible, but they are generally not as good for vegetable and fruit production as the more conventional types because their size is usually limited.

The basic frame of a greenhouse may be constructed of wood or metal—galvanized steel or aluminum, for example. Wooden

Metal-framed greenhouses are light and the frames are slender, allowing the maximum possible area of glazing.

Hoop houses offer lots of space for growing. They may not look sophisticated, but they make an excellent, functional choice.

frames look attractive, but the wood must be rot-resistant or it will warp and split. It must also be treated or painted fairly regularly to keep it in good condition.

Metal kit greenhouses generally need very little maintenance and are easier to construct, but the material retains less warmth than wood.

Although the glazing in a greenhouse is traditionally glass, it is possible to use polycarbonate or acrylic panels instead. This is cheaper than glass, but although it does not break, it scratches easily and heavily and needs renewing at regular intervals. Hoop houses (opposite) are usually made of polyethylene film stretched over a hooped PVC frame.

The floor of the greenhouse can consist of soil for beds, or may be poured concrete, lumber, or gravel throughout. Whichever you choose, the greenhouse walls must be erected on a solid, frost-proof foundation.

STAGING

If your greenhouse does not come equipped with staging, or shelving, you will need to install some to make the most of your space. Benches positioned along one wall at roughly waist height provide a working platform and a second level for growing, if you also use the floor or a low-level row of staging, too.

Staging can be made from wood or metal or wire shelving that also allow for drainage. Some staging can be reversible, with one solid side to act as a flat bench and one side with lips to form a tray that will retain water for irrigating plants left standing in it. Some are even deep enough to be used as a potting tray, or house removable plastic potting trays.

Adaptable systems make the staging versatile, allowing you to grow tall crops on the floor or lower level, lifting out the upper staging to give the crops more room.

Staging provides a comfortable working platform and leaves room beneath for storage, dormant plants, or more young seedlings.

POSITIONING THE GREENHOUSE

In most cases the greenhouse has to go wherever there is room for it, but it must be in a sunny position and not heavily overshadowed by trees.

It is generally most convenient if the greenhouse is positioned close to the house and it should have solid paths leading to it, to make it easier to reach with wheelbarrows and in wet or snowy weather. Some gardeners think that it is best to site a greenhouse with the roof ridge running east-west; others prefer a north-south orientation. East-west allows for more evenly distributed light to reach the whole greenhouse; north-south gives one warmer and one cooler side. Choose the orientation that suits your region, your site, and your intended crops.

Automatic window openers will help prevent overheating in the greenhouse, so growing plants do not "cook" on very hot days.

HEATING AND VENTILATION

All the greenhouse crops described in this book can be grown in a cold greenhouse. The only heat they will need is to encourage germination from seed, and this can be supplied by a propagator (*see* page 60). But in cold-winter areas, you will need to install some kind of heating.

For most commonly grown crops, greenhouse heating does not have to be the sophisticated equipment needed for growing orchids, for example. Simple heaters readily available in garden centers and home supply stores will do.

Choose the right fuel

Heating may be provided by gas, electricity, or even kerosene. Natural gas heaters are simple and efficient, but if they are to be run off the main house supply, the greenhouse must be close to the house, or the installation costs will be prohibitive. Propane heaters are stand-alone devices and work well for little cost. Those with a temperature control are the best. All gas heaters should be vented to the outside.

Electricity is probably the most efficient form of heating and you can choose between fan, convection, or portable radiators. Electric heating cables can also be buried in the soil to warm it up, but they will not help much to raise the overall temperature.

Allow air to circulate

Good ventilation is crucial to prevent mold developing and diseases thriving. At least two panes in the roof should be hinged so they can be opened and ideally there should be one or two along the sides of the greenhouse as well. Louvered panels allow you to provide slight ventilation, by just opening them a fraction.

Some greenhouses are equipped with thermostat-controlled openers that will open or close the panes according to the temperature inside. You can set your preferred opening and closing temperatures and leave them to operate automatically.

During summer, keep the windows open during the day and then close them at night if the outside temperature drops too low.

GREENHOUSE SHADING

Although greenhouses are designed to be warm, providing shading on very hot, sunny days is essential to prevent young plants, in particular, from getting scorched.

Shading can be provided by interior blinds or exterior shade cloth. Shade cloth must be supported by some kind of framework or secured to the framing with clips or grommets.

For glass greenhouses, a much cheaper alternative is to paint the glazing with whitewash or shading compound, which can be washed off with soapy water when it is no longer needed. Some shading compounds turn opaque in sunlight and fade back to being transparent when it is shady or overcast.

COLD FRAMES

Cold frames are invaluable for hardening off seedlings that have been raised indoors and would find the shock of going directly outdoors too great.

If the cold frame has a base of good soil, it may be used as a place to start seeds or to raise an early crop of lettuce or cabbage. Less hardy crops, like peppers or melons, can be grown in a cold frame from sowing to harvesting.

Cold frames come in just about as many different designs and sizes as greenhouses; they can be permanent structures attached to the side of the knee wall of a greenhouse (in which case they get heat from it), or they may be portable so you can put them virtually anywhere convenient.

Consider building your own

The basic cold frame design is a square or rectangular box about 6 in (15 cm) higher at the back than the front, with sloping sides. The frame itself may be constructed of wood, brick, or some sort of metal with glass or plastic sides, but the top must be glazed to allow light to penetrate. Ideally, the lid should be hinged at the back so it can be propped open on sunny days.

Many people make their own cold frames using an old window for the top and build the sides to fit the window frame so that they have the right size cold frame to suit their space. Anyone with a little basic woodworking experience should be able to build a simple frame to hold glass or polycarbonate for the top.

A window casement stay can be used to hold the lid open, or you could buy a special adjustable catch or just use a length of wood at varying angles to adjust the opening.

Position cold frames so that the sloping top is facing south, and keep the glass or plastic surface clean. Water plants sparingly so the atmosphere in the frame does not become too moist. You can keep the temperature stable by opening the top to varying degrees according to the weather. Use the cold frame to start seeds in spring, and to harden off crops raised in a greenhouse by gradually exposing them to longer periods of the cooler outside temperature before transplanting them into the vegetable bed.

Just as with a greenhouse, the size of the cold frame will depend on how much space you have available and how much you expect to use it. As a rule, aim for a minimum size of 4 ft x 3 ft (1.2 m x 1 m).

You can use cold frames to acclimate young plants to outdoor temperatures, so they are not shocked by a sudden move from indoors.

Greenhouse Management

A greenhouse provides an excellent growing environment for your crops, but it is also a fertile breeding ground for bacteria and diseases, and plants can easily fail as quickly as they first grew.

Hygiene and good maintenance are crucial to keep your greenhouse crops and young plants in good health. One of the simplest and most important preventive measures is to avoid overcrowding. When plants are kept too close together, moist, stagnant air collects and this will encourage disease. Keep an eye on seedlings as they grow and make sure that you thin them out promptly if they are growing in a propagator or seed tray, or spread them out if they are in individual pots. Pinch off any unnecessary foliage on established plants, such as some tomato leaves, to reduce the bulk of larger plants and allow air to circulate more freely.

SPOTTING SIGNS OF DISEASE

Open doors and windows in good weather to allow fresh air into the greenhouse and check all the plants in the greenhouse as you water them daily so that you are always vigilant for the first signs of disease.

Be scrupulous about removing diseased plants or leaves and taking them out of the greenhouse to dispose of them. If you spot signs of disease, treat it promptly and keep checking for indications that it is spreading. If necessary, be ruthless and get rid of any plants that look unhealthy—sacrificing one crop is better than losing all the plants growing in the greenhouse.

Don't allow dead leaves, garbage, or spilled potting soil to linger in the greenhouse. A paved floor is a good choice because it is so easy to sweep clean. If you have soil beds in the greenhouse, weed them regularly and pick out fallen leaves or other bits of plant debris left on the surface.

WATERING AND DAMPING DOWN

Although watering is vital, it is easy to overdo it and leave plants waterlogged. A little and often is far better than days of drought followed by a drenching. Water daily during the growing season and preferably twice on the hottest days.

Always use a water can with a wide rose or a wide spray from a hose-end sprayer to gently shower water over seedlings and young plants to avoid damaging them with a deluge.

Another way to get much-needed moisture to your plants in the heat of summer is to increase the humidity in the greenhouse. You can do this easily by spraying the leaves of your crops with a plant mister, rather than only watering at the roots.

In addition, you can use a gentle spray from the hose to "damp down" the greenhouse floor, if it is paved. The heat of the paving and of the sun will make the water evaporate.

FALL AND SPRING CLEANING

Some diseases will inevitably find their way into your greenhouse and spores and bacteria can lie dormant in the fabric of the building from season to season, ready to attack your crops next year.

In fall, when the summer crops have finished and you are no longer raising seedlings, give the greenhouse a really thorough cleaning. Choose a mild day so you can move any plants that are ready to overwinter under glass outside while you work. Scrub the outside of any containers that will go back into the greenhouse later, taking care not to damage the plants in them.

Clear the greenhouse of all empty pots, potting equipment, labels, and tools. Clean them all thoroughly, paying attention to the rims of pots and any corners, cleaning off moss and scraps of potting soil. Dry and store them neatly, ready for next year. Scrub seed trays with a mild bleach solution and allow them to dry thoroughly before storing.

Next, sweep the floor and brush the frame, staging, and glazing to get rid of any cobwebs, dust, or debris. Make a note of any damaged panes of glass or parts of the framework that need attention and fix them as soon as possible.

Wash the glass inside and out using a soft cloth and a bucket of warm water mixed with mild dishwashing liquid.

Scrub the frame and staging with a stiff brush, and clean mold and algae from between overlapping panes of glass by sliding a thin plant label or piece of plastic between the panes and moving it up and down. Remove the glazing to clean it thoroughly if necessary. The manufacturer may have special cleaning solutions for acrylic or polycarbonate glazing.

Rinse everything with strong blasts of water from the hose, or even a pressure washer, to remove any soapy residue. Remember to include the floor and the underside of all shelving.

It may seem laborious, but cleaning the greenhouse thoroughly will save you a good deal of time and effort next growing season in tackling pests and diseases on your crops.

Cloches

You can speed up plant growth by creating warmer, more humid conditions, and cloches are the simplest way of doing this. Use them to raise the temperature of the soil before sowing seed and to protect vulnerable seedlings.

Hoop houses cover whole rows of seedlings. They are efficient and economical, whether made from greenhouse plastic or row covers.

Cloches have evolved from a sort of glass dome that was placed over one or more plants. They are now most commonly used as hoop houses to cover a row of germinating seeds, young seedlings, or even mature plants such as strawberries, to ripen them more quickly. They are also useful when placed over the soil in spring to warm it prior to sowing.

Hoop houses are often made of flexible PVC hoops covered with row covers or greenhouse plastic sheeting, which you can find in different weights from greenhouse and landscape supply companies. A common method of construction is to build a simple base of lumber and secure the PVC pipes to the base using clips or straps. Netting may also be draped over the hoops to protect the bed from birds and deer.

Provide some fresh air inside

Glass cloches are still regarded by many to be the best; they let in the maximum amount of light and do not cause condensation (which plastic will do), but they are expensive, breakable, and often difficult to get hold of in garden centers.

Whatever type of cloche you use, it must be closed at the ends if it is to be fully effective. But when plants are growing, closed cloches can become stale and condensation build up inside. Some plastic cloches now come with ventilation holes that can be opened and closed as required to reduce the condensation.

If your cloches are open-ended (this makes them useful for joining together to cover a long row) close the ends with a pane of glass or sheet of plastic held in place with a rock, otherwise all the warm air under the cover will escape through the open ends. You can tie the ends of sheets of plastic to close them then stake them into the ground.

Keep an eye on the plants

Remember to check under covers regularly. The plants inside may need watering more frequently than those in the open soil, as the warm ground will dry out faster. And weeds will thrive under a protective covering just as well as your crops.

If you are using cloches to protect frost-tender plants, such as strawberries, remember to lift or open them to allow insects access to the plants when they need pollinating. You can do this by removing one rigid cloche and spacing the others out, or by lifting one side of the plastic sheeting.

Traditional glass "bell" cloches (front) or larger, square lantern cloches (behind) are an attractive way of protecting tender young plants in the vegetable garden, but they are only suitable for individual plants, so are of limited use to the self-sufficient grower.

Crop Rotation

One of the first principles of growing vegetables is that the crops in each plant group should be grown in a different part of the vegetable garden each year, if possible.

This program of crop rotation is important for two main reasons: the first is that different crops like different soil conditions. Legumes (peas and beans), for example, like ground that is rich in freshly added, well-rotted manure, but if root vegetables such as carrots and parsnips are grown in the same soil, the result is twisted, forked, divided vegetables that have gone on a search for organic matter.

The second reason is that the pests and diseases that plague specific families of vegetables will remain in the ground, ready to attack again. They do not, however, affect vegetables in the other groups.

A typical crop rotation operates on a three-year plan, although if you have enough space, you can divide the garden up into four areas and practice a four-year crop rotation.

CROP ROTATION PLAN

It is vital to rotate your vegetable crops, because different crops like different soil conditions and extract different nutrients from the soil. Also, pests and diseases will attack one type of crop but leave another unaffected.

DEVISING YOUR CROP ROTATION PLAN

Following the three-year rotation plan, you must divide your garden into three areas and your crops into the three principal groups. These groups are the root crops (*see* page 66, but also including potatoes and endive), cole crops (*see* page 62, plus radishes), and a combination of the legumes and other outdoor crops that, in addition to beans and peas, include celery, celeriac, garlic, leeks, onions, corn, peppers, and tomatoes.

Salad greens and other fast-growing crops not mentioned in these groups can be grown as catch crops—that is, planted in beds after they have been cleared of a major crop and before the next one is sown—or interplanted between rows of slower-growing vegetables.

Squash, zucchini, melon, and pumpkins can be included in the legume and other outdoor crop section, but if you have a piece of spare land or a space on a compost pile, you can grown them there, for they are sprawling plants that can take up valuable space in the vegetable garden.

The perennial crops, such as artichokes and asparagus, have to be given a permanent space as they will crop in the same site for a number of years. Jerusalem artichokes can be included in the root vegetable group, but they spread so vigorously that they are really best grown on a small patch of otherwise unused ground, such as in front of the garden shed or positioned to hide the compost pile and other work areas from view. If you intend to operate a three-year rotation plan, and have divided the main part of the vegetable garden into three roughly equal-sized beds, allocate the first bed to the root crops, the second bed to the legumes, and other outdoor crops, and the third bed to the cole crops. The following year move the vegetables down the line, so that the first plot has the cole crops, the second has the root crops, and the third the legumes and other outdoor crops. In the final year of the plan, the first plot will have the legumes, the second will have the cole crops, and the third will have the roots.

Constructing equal-sized raised beds is a good way to mark out zones for a crop rotation plan. Grow one whole bed of cole crops, another of root vegetables, and a third of legumes.

Feeding the soil

You will get even better crops if you follow a system of digging lots of well-rotted manure or compost into the ground before planting the legumes, applying a balanced fertilizer to the ground before planting root vegetables, and applying both fertilizer and lime before planting cole crops.

The roots will benefit from the manure dug in for the legumes, because by the time they are planted it will have become well-rotted and incorporated into the ground, thus making the soil moisture-retentive, rather than very rich. Potatoes, which do not like an alkaline soil, are not planted until two years after soil has been limed, but cole crops, which thrive in alkaline conditions, will flourish there.

A four-year rotation plan

If you are able to operate a four-year plan you can either separate out the peas and beans from the onions and other outdoor crops, or you can treat potatoes as a separate group, perhaps using that bed also for squash and zucchini and some of the leafy greens, such as spinach or Swiss chard. Divide your land into four equal areas and move crops one plot along each year.

GOLDEN RULES FOR SUCCESS

The most important things to remember in any rotation plan are to ensure that cole crops and potatoes never follow themselves in the same sites in successive years (and preferably not more than once every three years) and to plant quick-growing, successive crops after harvesting a major crop whenever possible.

If cabbages follow Brussels sprouts, you could easily get a build-up of the disease clubroot (*see* page 148) in the soil and once there, it will remain for several years, making it impossible to grow any cole crops at all. Likewise with potatoes, you could encourage nematodes or blight (*see* page 151) in the soil, and these problems, too, will remain for some time to infect further crops.

Remember, that some crops that take time to reach full maturity can be regarded as quick-growing when grown as successive crops. Carrots and beets, for example, which would take perhaps four months to reach full size, can be harvested much quicker as young and delicious baby vegetables. In some instances you will find that there is time to grow a quick-maturing crop after harvesting the previous one, and before planting the next one you have lined up in your crop rotation.

Watering and Irrigation

In many parts of North America, drought is either a seasonal or a year-round reality. There are ways to minimize the amount of watering you need to do, and to conserve supplies of clean, treated tap water.

Watering with a hose is quick and easy—with a sprinkler attached it's even easier—and it is tempting to do this when you have a large area to water and so many other tasks calling for your attention. But a sprinkler will use many gallons of water per minute, so that running a sprinkler for an hour can use hundreds of gallons of water.

As our climate changes, water is becoming a more scarce and valuable commodity at the same time as it becomes more essential for the self-sufficient homesteader. You can do a lot to conserve water in the choices you make about how you use it, the plants and varieties you choose to grow, and how you design and maintain your garden.

Follow the advice on these pages to cut down the amount of water you use and you can also save yourself a considerable amount of money as well as helping to conserve precious supplies of water and the energy used to treat it and deliver it.

Direct water carefully to the base of each plant to make the most of each drop; widespread sprinkling is wasteful and ineffective.

WHEN AND HOW TO WATER

Make the most of the water you give to your plants by targeting it well and watering at the right time of day. Always try to avoid watering plants at the hottest part of the day. The water will evaporate fast in the heat and much of it will be wasted.

Instead, water at the beginning or end of the day. Water at the base of each plant so that the water goes directly to the roots. With sprawling crops, such as pumpkins, it is a good idea to push a stake into the ground where the stalk emerges to help you to direct your water to the right spot.

Particularly thirsty crops—again, pumpkins are a good example—can benefit from a watering well. Build up a raised lip of soil around the stem at a diameter of around 18 in (50 cm); this will retain a well of water that will seep slowly into the soil, giving the plant a really good soaking that will last it for a couple of days in mild weather.

Walking around with a hose fitted with a hose-end sprayer to water your garden offers the opportunity to check plants for signs of pests or disease as you go. You can use a strong spray of water from the hose to dislodge pests such as aphids from your crops. Use a gentle spray setting when watering young seedlings so that you don't disturb the soil around them.

One golden rule of watering is not to water plants unnecessarily. In the open ground, where plants have access to ground water stored in the soil, you may only need to water every couple of days in average temperatures, especially in spring and early summer. Be alert for signs of drought—wilting foliage and leaves that are starting to look crisp and sunburned—and increase your watering regime for a while until the weather cools again.

How soil holds water

If your soil is heavy and prone to waterlogging, you will see puddles sitting on the surface after heavy rain. You can dig in gypsum to improve it or install a drainage system (*see* page 27). Light sandy soil will drain very quickly, sometimes before plants can benefit from the water you give them, but this is harder to identify. To gauge how well your soil retains water, dig a large hole and fill it with water, then monitor how long it takes for the water to drain away. If the hole empties within an hour, plants are unlikely to thrive in the soil; if the water is all gone within four hours, the soil would benefit from more organic matter. Dig in plenty of well-rotted compost and you should find that you need to water less and less often.

Automatic irrigation systems like this can be useful in a very large garden, where watering by hand is not practical. Set the timer to make efficient use of your irrigation water.

AUTOMATIC IRRIGATION SYSTEMS

Almost any vegetable and fruit garden can benefit from the installation of an automatic irrigation system. It conserves water, and ensures that your plants are watered only as much as they require.

While a lot of the water from a sprinkler will inevitably fall on bare earth, paths, and other areas that do not actually require watering, automatic irrigation systems are more precise. You can have a system installed by a professional, or you can install a simple drip irrigation system yourself.

Your choices are to run over-ground flexible ½ in (12 mm) irrigation tubing across the surface of the ground, or to dig trenches for rigid PVC pipe that deliver water to above-ground sprinklers or drip irrigation tubing. The result, though, delivers water directly to each plant or to a small area at a time.

Electronic timers and valves control how long the water runs and when it is switched on, and the most sophisticated systems allow you to set the timer for different sections of garden, delivering more water where it is needed and a light sprinkling to less thirsty crops.

Soaker hoses can also be laid on top of the ground through the vegetable garden. These deliver water as it seeps slowly through the fabric of the hose along its entire length, giving a gentle, slow watering to a strip of land. Cover the soaker hose with mulch to further conserve moisture in the soil.

COLLECTING AND USING RAINWATER

The very best water to use for your plants is rainwater. It is free, kept at the same temperature as the air—rather than the chill of tap water—and contains none of the chemicals found in treated water.

Position rain barrels or tanks around your property wherever you have downspouts or any pitched roofs (*see* page 247). Rain barrels are available in various sizes, typically 55 or 90 gal (200 or 350 liters). You could be surprised at how quickly a rain barrel will fill up, even if it is only draining off a small shed. If you have room, you can join several barrels together to collect more water. Even if you don't have any suitable buildings on one part of your property you could stand an open tank in an unused space and collect water that falls into it. Make sure that you cover the top with hardware cloth, preferably fixed securely in place, to prevent animals or children falling in and drowning.

Fill waterings cans from the spout at the base of the barrel or, to water the garden quickly, attach a garden hose to a submersible pump lowered into the barrel. Gravity-fed rainwater systems are also available, which pipe water from the barrel to sprinklers or drip irrigation tubing, similar to those found on other automatic irrigation systems.

Rain water collected and stored in rain barrels is perfect for plants: it does not contain chemicals and is not delivered at high pressure and low temperature, like tap water.

PLANNING A WATERWISE GARDEN

You can help to protect your plants from heat and drought by designing your garden carefully and by choosing drought-resistant varieties if you live in an area that has long, dry spells and hot summers.

One of the most important things you can do before you start planting anything is to check the water-retaining quality of your soil (*see* page 42). If there is insufficient humus mixed in with the soil particles, water will drain through too freely and much of it will be wasted.

Trees and hedges are greedy and thirsty plants, and will suck up large quantities of water and nutrients in the soil, so if you position your vegetable garden too close you will increase your plants' need for extra feed and water, making more work for yourself as well as extra demands on your water supply.

Protect against sun and wind

Overhanging trees will also cast shade over your garden, as will a hedge, fence, or any other solid boundary unless it is to the north of the bed. While some shade can help to reduce wilting in very hot sun, in cooler areas it is likely to make it more difficult for your plants to thrive and, again, you will need to work harder to achieve success than in a bright, sunny situation. Choose a site that gets at least six hours of sun a day—preferably in the morning, rather than the afternoon and evening sun.

In regions that often have periods of hot, dry, sunny weather, consider erecting a simple pergola over an crops that might be susceptible to drought, such as strawberries or lettuce. The wooden framework will provide light shade without blocking out much-needed sun at other times of year. It can be draped with some shade cloth on days when you need to add more protection.

Persistent wind will reduce the water-retention of your soil, so if you cannot avoid planning your garden in an exposed site that you know will be subject to high winds, work windbreaks into your plan. These could be trellis panels that can help to minimize the effects of the wind without blocking out too much sun, or could be living windbreaks, such as berries or artichokes planted along one edge of the garden.

When you plant your plot, try to avoid leaving large areas of bare dirt close to planted sections, because water can evaporate from the bare ground in the sun or be stripped away by winds, and this will take valuable moisture away from the plants that are growing nearby. Use a mulch of compost or tree bark between rows and on empty patches of your garden to prevent this. A generous mulch of organic matter will also rot down into the soil, feeding it at the same time.

Green manures

If you have a large area of ground that is lying fallow, consider planting a green manure. These are fast-growing crops, often rye grass, mustard, vetch, or peas that cover the ground for a season, often over winter. When they have died down, you should dig the foliage into the soil, where it will enrich it with nutrients as it rots. Garden centers and mail-order suppliers supply a range of green manure crops, many

Green manures will cover an area of bare soil, helping to retain moisture. They will feed the soil, too, when you dig them in later.

of them tailored to the particular needs of different types of soil and different regions of the country.

Test your soil (*see* page 24) to find out more about its structure and nutrient levels, then choose an appropriate green manure crop.

CHOOSING THE RIGHT PLANTS

Plant breeders are all aware of the need to use less water and are constantly working to develop new varieties that are increasingly drought-tolerant. Look for types of vegetables and some specific varieties that are better able to withstand hot, dry conditions.

Root crops, such as carrots, are generally more tolerant of drought than plants with swelling fruits, such as zucchini and tomatoes. Root crops can penetrate deep into the soil to reach moisture locked below the surface. Leeks also survive well in dry conditions; hilling up (a process of scraping soil around the stems to blanch them) also helps to lock in moisture. Leafy crops are less forgiving of drought, but Swiss chard and mustard greens are more drought-tolerant than other greens.

Onions, arugula, and lettuce are prone to bolting, or running to seed, in hot, dry weather, so look for drought-resistant varieties. In general, try to choose plants that grow in naturally hot regions if your own area is often affected by high temperatures. Okra, chili peppers, and sweet potatoes, for example, which once were possible only in the South, are now widely grown.

CALENDAR OF SEASONAL TASKS

The Year in the Garden

The calendar of tasks outlined on the following pages is meant only as a general guide to the main chores of sowing, planting, and harvesting. With a huge variety of climates and microclimates across North America, you will have to make adjustments.

The climate in your region will dictate the actual timing for doing these tasks, and this will vary from year to year. This calendar follows the seasons rather than months, because you could safely start planting out in the South months before more northern areas, where temperatures are lower and winters usually colder and longer.

Be guided by the weather conditions and signs around you in nature to gauge when it is warm enough in spring to start sowing (buds swelling on trees and emerging bulbs, for example); and keep an eye on a maximum/minimum thermometer to spot changes in temperatures. A knowledgeable local nursery and your local Cooperative Extension Office are also good sources for determining when to plant and harvest in your area.

Understanding your garden and animals takes time and experience, but keeping records can help. Note down the weather, what you have done, and anything that was ready for harvesting, transplanting, sowing, pruning, and more. Compare your notes the following year and you will soon start to get a feel for the rhythm of the seasons and the cycles of nature.

Tailor the calendar to your needs
There are many crops that can be sown in regular succession from spring to fall, but if you do this you can be overwhelmed with that one vegetable and may not have much room left in the vegetable garden to grow anything else. Sow all vegetables as and when it suits you (within their particular season), to get a reasonable amount and assortment of vegetables, rather than an ongoing glut of any one type.

Routine garden tasks
Remember that, in addition to the jobs described in this calendar, you must weed the fruit and vegetable garden regularly throughout the growing season to keep weeds under control, and all crops must be watered regularly in dry weather. Stake and tie plants as necessary to support them as they are growing, and protect emerging seedlings from birds by tying threads of cotton above them or by covering them with row covers. In addition, protect young crops from frost in the winter and early spring with cloches.

Inspect all crops regularly to make sure they are not suffering from any disease or pest infestation; if they are, deal with it immediately (*see* page 143) to prevent the condition getting worse or spreading, and dispose of any infected plants you pull up. Dig up and compost all healthy plants once the harvest is finished, so that the bed can be used for another crop. Try not to allow any part of the vegetable garden to lie fallow, particularly during the main growing season.

Spring

As bulbs start emerging in the garden and buds begin to swell on the trees, it is time to venture back into the vegetable and fruit garden.

You may be able to harvest some early spring greens or rhubarb and overwintered root crops, but keep an eye on the weather and watch out for sudden spring frosts that can strip the blossom from fruit trees and destroy your budding crop.

As the season progresses and the soil warms up, seedlings, blossoms, insects, and the promise of a bountiful harvest are all around you.

EARLY SPRING

In the greenhouse

- Water all greenhouse crops regularly throughout the spring
- Sow peppers, eggplants, and tomato seeds and raise them in a propagator

In the fruit garden

- Plant apricots, figs, vines, and rhubarb
- Sow seeds of strawberries under glass or indoors
- Prune apples, apricots, cherries, and European and Japanese plums
- Pollinate apricots
- Raise new plants of rhubarb
- Harvest rhubarb

In the vegetable garden

- Prepare raised beds and permanent growing sites for sowing as soon as the weather conditions allow
- Sow under glass or indoors seeds of globe varieties of beets, artichokes, asparagus beans, broccoli, leeks, self-blanching celery, green-stemmed celery, and onions
- Sow in permanent sites seeds of kohlrabi, Asian greens, broad beans, maincrop carrots, parsnips, salsify and black salsify, early peas, snow peas, petit pois, lettuce, mesclun, radishes, scallions, onions from seed, and summer spinach
- Plant early potatoes and Jerusalem artichoke tubers
- Harden off summer cabbage and cauliflower
- Harvest overwintered broccoli, cabbage, kale, perpetual spinach, rutabagas, turnips, and early greens, endive, lettuce, and winter spinach
- Protect early crops and young seedlings from cold weather with cloches as necessary

Follow the weather forecast and protect outdoor crops with floating row covers or portable hoop houses when frosts are expected so that they are not lost early.

In the herb bed

- Sow in permanent sites seeds of burnet, chervil, fennel, dill, lovage, marjoram, parsley, rosemary, sage, and thyme
- Divide roots of lovage to create new plants
- Plant bay trees, lavender bushes, and tarragon
- Start basil indoors

Get an early start with tomato seedlings sown and raised in a propagator in the greenhouse. Move them outside as the weather warms up.

MIDSPRING

In the greenhouse

- Transplant eggplant and cucumber seedlings to their permanent growing positions
- Sow melon seeds and raise them in a propagator
- Prune vines and established fruit trees. Train new growth on wires fixed to the greenhouse frame
- Pinch out growing tips of vines
- Start opening vents to increase ventilation as the daytime temperatures rise
- Damp down the greenhouse floor each time you water the plants to increase humidity

In the fruit garden

- Harden off strawberries
- Train new growth of vines, peaches, and nectarines
- Mulch apples, apricots, sour cherries, peaches and nectarines, pears, gooseberries, vines, raspberries, and rhubarb. Put straw under strawberry plants
- Harvest rhubarb

In the vegetable garden

- Sow indoors: pole and bush beans, tomatoes for outdoors, celery, celeriac, squash, zucchini, pumpkins, okra, and New Zealand spinach
- Sow in temporary beds seeds of broccoli, Brussels sprouts, summer cabbage, and kale
- Sow in permanent sites seeds of kohlrabi, maincrop carrots, parsnips, salsify and black salsify, perpetual spinach, turnips, peas, snow peas, petit pois, endive, lettuce, radish, leeks, scallions, onions from seed, onions from sets, garlic, and summer spinach
- Plant second-early potatoes and artichoke crowns
- Harden off broccoli, globe varieties of beets, asparagus beans, tomatoes for outdoors, celery, celeriac, leeks, and onions from seed
- Transplant to their permanent sites globe varieties of beet, carrots, and turnips
- Harvest early broccoli, spring cabbage, savoy, red cabbage, kale, perpetual spinach, rutabagas, turnip greens, lettuce, onions, leeks, and winter spinach
- Begin earthing up emerging potatoes

In the herb bed

- Sow angelica, balm, borage, burnet, chervil, coriander, dill hyssop, pot marjoram, sage, and summer savory

LATE SPRING

In the greenhouse

- Transplant pepper, melon, and tomato seedlings to their permanent sites in greenhouse beds, containers, or grow bags
- Mulch all greenhouse crops
- Shade greenhouse roof as necessary with shading paint, blind, or shadecloth
- Pollinate fruit trees under glass
- Support growing eggplant and cucumber plants

In the fruit garden

- Transplant strawberries to permanent beds
- Thin fruits of apricots and gooseberries
- Harvest rhubarb
- Fertilize and mulch blueberries
- In warm climates, plant lemons, limes, oranges, and other citrus
- Train growing vines such as grape and kiwi, so the shoots do not become tangled and unruly

In the vegetable garden

- Sow indoors seeds of warm-season crops like cucumber, celery, and celeriac
- Sow in temporary seed beds Brussels sprouts, cabbage, cauliflower, broccoli, and broccoli raab
- Sow in permanent sites seeds of kohlrabi, maincrop carrots, salsify and black salsify, Swiss chard, rutabagas, turnips, kale, broad beans, bush and pole beans, green beans, late peas, snow peas, petit pois, cucumber (protect with cloches in coldest areas), onions from seed, okra (but protect with cloches in coldest area), and New Zealand spinach
- Sow corn in blocks, at two week intervals
- Plant maincrop potatoes, asparagus crowns
- Harden off beans, green beans, squash, zucchini, pumpkins, okra, melon, New Zealand spinach, and celery
- Transplant beans, asparagus beans, eggplant, cauliflower, celery, celeriac, onions from seed, and okra
- Thin carrots, parsnips, turnips, dwarf, beans, lettuce, arugula, and all types of spinach
- Harvest broccoli, spring cabbage, rutabagas, turnip greens, broad beans, early peas, asparagus, lettuce, radish, onions, scallions, leeks, and winter spinach
- Continue hilling up potatoes

In the herb bed

- Sow chervil, dill, pot marjoram, and sage
- Harden off seedlings of basil and marjoram grown under glass and plant out into containers or the herb bed a few weeks later

Summer

The fruit and vegetable garden should, by now, be bursting with produce and growing fast. Summer is a season of hard work, but rich rewards—what could be better than collecting a basketful of fresh and delicious produce for your evening meal?

You will be busy throughout the fruit and vegetable garden in summer, watering, thinning, and harvesting plants and checking thoroughly and regularly for signs of pests or disease. Feed plants regularly, too, to get the best out of them.

Check your crops daily, even if they don't require daily watering, so that you can spot and pick produce when it is at its very best. A young zucchini can very quickly turn into a monster, losing its delicious flavor overnight and becoming tough-skinned and watery; and soft fruits and tomatoes can overripen and begin to rot on the bush if they are not harvested promptly.

Preserving produce

At this time of year you may find yourself with a glut of one or more crops that have all matured at once. This spare produce can often be used to repay favors or store up future goodwill with friends and neighbors or you may be able to trade your surplus with another grower's different glut. But anything you cannot eat, swap, or give away should be preserved to see you through the leaner winter months.

Many things—soft fruits in particular—can be frozen in their natural, harvested state or, with beans, for example, blanched quickly in boiling water first.

Freeze your produce on open trays to prevent individual berries or vegetables clumping together in a mass that must all be defrosted and used at once (*see* pages 210–11 and 222–23 for more on freezing).

The vegetable garden in summer is bursting with produce and there is daily work to be done, harvesting, thinning, and watering.

Other crops can be preserved by cooking and then freezing or bottling. Turn a basket full of tomatoes into a pasta sauce, tomato ketchup, chutney, or the basis of a ratatouille with leeks, zucchini, and peppers. Or make jams, jellies, pickles, and chutneys (*see* pages 214–17 and 226–29) that will last for months in a cool cupboard or pantry.

Carrots, potatoes, onions, apples, and other crops can be stored for several months in a cool place if they are cleaned and packed carefully (*see* pages 207–08). And some root crops are actually best left in the ground if you do not need the space immediately for another crop.

Don't forget that some slightly spoiled produce that doesn't get picked in time or that falls from the bush before you pick it can still be used in chutneys or may be fed to your pigs. And anything that you miss until it is really too late for consumption should be added to the compost pile.

EARLY SUMMER

In the greenhouse

- Throughout the summer, water and feed all greenhouse crops regularly—twice a day, in morning and early evening, if necessary
- Pollinate grapes and melons
- Thin out tree fruits
- Support growing peppers, melons, and tomato plants
- Harvest cucumbers
- Apply greenhouse shading to the glass if not already done in spring to protect plants from scorching

In the fruit garden

- Prune peaches and nectarines, gooseberries, and red and white currants
- Thin fruits of apples, plums, peaches, nectarines, and gooseberries
- Harvest strawberries, black currants, gooseberries, and rhubarb

In the vegetable garden

- Water all crops regularly during dry spells
- Support growing plants
- Sow in temporary beds seeds of red cabbage
- Sow in permanent sites seeds of Chinese cabbage, kohlrabi, globe varieties of beets, long-rooted varieties of beets, pole and bush beans, first-early peas, asparagus beans, lettuce, radish, and onions
- Harden off cucumbers
- Transplant to permanent sites broccoli raab, Brussels sprouts, fall and winter cabbage, savoy, cauliflower and kale, green beans, tomatoes for outdoors, cucumbers, and pumpkins
- Thin Chinese cabbage, kohlrabi, salsify and black salsify, summer spinach, New Zealand spinach, Swiss chard, and perpetual spinach; use the thinnings in salads and stirfries
- Pinch out growing tips of trailing squash and pumpkins
- Harvest artichokes, summer cabbage, kohlrabi, beets, carrots, broad beans, pole and bush beans, first-early peas, snow peas and petit pois, asparagus beans, early potatoes, asparagus, land cress, lettuce, radish, salad onions, summer spinach, and Asian greens

Pick perpetual spinach and leafy greens, like this ruby chard, regularly to encourage new growth of tender young stems.

In the herb bed

- Sow in seed beds seeds of chervil and dill
- Harvest herbs as required for cooking

MIDSUMMER

In the greenhouse

- Thin out fruits of grapes and melons
- Harvest eggplants, cucumbers, tomatoes, peppers, chili peppers, and tree fruits
- Water and feed crops regularly, looking out for signs of drought
- Inspect crops daily for signs of disease and treat promptly

In the fruit garden

- Plant strawberries
- Prune apples, peaches and nectarines, and pears
- Thin fruits of pears (if necessary) and grapes
- Train blackberries, loganberries, and raspberries
- Raise new plants of strawberries from runners
- Harvest apricots, sour cherries, peaches and nectarines, alpine strawberries, strawberries, black currants, blueberries,

gooseberries, summer-fruiting raspberries, red and white currants, and rhubarb

In the vegetable garden

- Sow in temporary beds the seeds of red cabbage
- Sow in permanent sites kohlrabi, globe beets, perpetual spinach, turnips, beans, onions, scallions, and summer spinach
- Continue to sow greens like mache, lettuce, Asian greens, arugula, mustard, amaranth, and mizuna
- Transplant to permanent sites broccoli raab, Brussels sprouts, fall and winter cabbage, savoy, winter cauliflower, and kale
- Thin kale, kohlrabi, beets (all varieties), parsnips, salsify and black salsify, endive, summer spinach, New Zealand spinach, and rutabagas
- Pinch out growing tips of broad beans, tomatoes, and New Zealand spinach
- Pollinate cucumbers, pumpkins, and squashes
- Harvest artichokes, summer cabbage, kohlrabi, beets, carrots, turnips, broad beans, pole and bush beans, green beans, first-early peas, second-early peas, maincrop peas, snow peas and petit pois, asparagus beans, early- and second-early potatoes, land cress, lettuce, radish, garlic, squash and zucchini, onions, scallions, summer spinach, and New Zealand spinach
- Shade lettuce and other greens with floating row covers in hot weather to prevent them from bolting
- Keep plants well-irrigated

A cherry tree is a froth of wonderful blossom in spring and will give a good crop of fruit in summer. Pick the cherries as soon as they are ripe.

In the herb bed

- Sow in permanent sites seeds of chervil, dill, and parsley
- Pinch out growing tips of bushy herbs, such as basil and marjoram to encourage growth
- Watch for herbs like coriander and parsley bolting and pinch out the tips of plants that start to bolt

LATE SUMMER

In the greenhouse

- Harvest eggplants, cucumbers, tomatoes, peppers and chili peppers, and tree fruits

In the fruit garden

- Continue harvesting as fruits ripen
- Cut back June-bearing raspberry canes to just above ground level after fruiting
- Prune and train vines growing outdoors
- Raise new plants of strawberries by layering runners

In the vegetable garden

- Sow in temporary beds seeds of red cabbage, and lettuce, and other fall and winter greens
- Sow in permanent sites seeds of kohlrabi, perpetual spinach, rutabagas, garlic, turnips, turnip greens, mache, endive, lettuce, radish, winter radish, onions, scallions, onions from seed (overwintering varieties), and summer and winter spinach
- Transplant to permanent sites cauliflower, broccoli, and summer cabbage
- Thin Chinese cabbage and tie up outer leaves to encourage inner growth
- Thin kohlrabi, endive, spinach, and perpetual spinach
- Blanch celery
- Harvest summer cabbage, kohlrabi, beets, carrots, Swiss chard and perpetual spinach, rutabagas, turnips, broad beans, pole and bush beans, green beans, first-early peas, maincrop peas, late peas, snow peas and petit pois, asparagus beans, second-early potatoes, outdoor tomatoes, land cress, lettuce, radish, garlic, marrows and zucchini, onions, shallots, scallions, okra, summer spinach, New Zealand spinach, and corn
- Continue to keep plants well-irrigated and cover with shade cloth or row covers to prevent plants scorching in hot climates

In the herb bed

- Sow chervil, marjoram, and thyme

Fall

Season of the traditional harvest festival, fall—at its most crisp and clear—is one of the most pleasurable times of year to be in the garden. Enjoy the late gathering in of produce at the close of the growing year. In warm-winter climates, now is the time to plant and sow for winter harvests.

In cool-winter areas, it's time to start preparing the garden for winter. Except for winter varieties of cabbage and cauliflower, Brussels sprouts, and other winter greens, most green, leafy crops must be harvested before the first of the frosts ruins them. As your food mood starts to turn to warming casseroles and stews, pull up turnips, carrots, and rutabagas.

Windfalls of apples and pears can be collected and used for cooking—or leave some on the ground for the birds as the season draws to a close. To avoid your crop spoiling, pick the fruit from the trees and wrap and store each one carefully.

Choose a fine day to clear and clean your greenhouse and to scrub off any shading paint applied for the summer. On another fair day, dig over any bare ground and sow a green manure or spread on a layer of well-rotted manure or compost to enrich the soil over the winter.

EARLY FALL

In the greenhouse

- Water and spray all greenhouse crops regularly
- Harvest eggplant, cucumbers, tomatoes, tree fruits, and grapes
- Clean the greenhouse of debris and wash down all surfaces

In the fruit garden

- Prepare the ground for planting tree fruits and canes
- Mulch plums and black currants
- Thin fruits of figs
- Raise new plants of black currants, gooseberries, and other currants
- Harvest apples, figs, pears, strawberries, alpine strawberries, blackberries and loganberries, blueberries, grapes, and everbearing raspberries
- Continue to harvest citrus fruits and check for pests

In the vegetable garden

- Sow indoors seeds of cauliflower and lettuce for overwintering and forcing
- Sow in permanent sites mache, cress, Asian greens, kale, and winter spinach
- Thin mache, salad greens, and winter radish
- Harvest broccoli, early Brussels sprouts, cabbages, cauliflower, kohlrabi, beets, carrots, parsnips, Swiss chard and perpetual spinach, rutabagas, turnips, dwarf, pole and bush beans, green beans, late peas, snow peas and petit pois, lettuce, radish, celery, zucchini and pumpkins, okra, summer spinach, New Zealand spinach, onions, and corn

In the herb bed

- Sow angelica and lovage
- Take root cuttings of sage and grow indoors
- Harvest seedheads of fennel for drying
- Harvest herbs for drying like basil, rosemary, sage, and thyme, before fall and winter rains start in earnest

LATE FALL

In the greenhouse

- Harvest eggplants, tomatoes, tree fruits, and grapes
- Pull up and compost all annual plants when harvesting is finished
- Prune fruit trees and vines
- Clean and disinfect (see page 38), including cold frames, cloches, containers, and trays
- Insulate if necessary to protect overwintering tender crops

In the fruit garden

- Plant apples, apricots, plums, peaches and nectarines, blackberries, loganberries, black currants, gooseberries, and red and white currants
- Prune black currants and raspberries
- Protect strawberries with hoop houses or floating row covers to give a final late crop
- Harvest apples, figs, grapes, pears, strawberries, alpine strawberries, and everbearing raspberries
- Lift roots of rhubarb for early forcing (see early winter)

In the vegetable garden

- Sow indoors over-wintering or forcing lettuce
- Sow in permanent sites broad beans (longpod) and garlic
- Thin mache and lettuce
- Blanch endive and force endive
- Protect red cabbage with cloches and cauliflower curds by bending over the outside leaves
- Cut back asparagus foliage
- Harvest broccoli, broccoli raab, Brussels sprouts, perpetual spinach, cabbage, cauliflower, kale, kohlrabi, beets, carrots, parsnips, salsify, chard, cress, rutabagas, turnips, lettuce, and leeks

Winter

In winter, when there is work to do in garden, it is a good time to look back and reflect on the successes and failures of the year and to plan the next year's crops.

Make the most of nice days to finish pulling up, pruning, and composting or discarding your plants. Dig over bare dirt in early winter to allow frosts to kill any pests that come to the surface and to break down compacted clumps of dirt. By spring, you will just need to rake it. Clean your tools and oil any moving parts before putting them away.

Harvest root crops by the end of fall and store them carefully in a cool place; freezing winter conditions will spoil them in the ground.

On days when the weather keeps you from all but the most pressing jobs outside, thumb through seed catalog and the pages of this book to decide what to grow next year and start to make a plan so that you are ready in spring. As winter draws to a close you can even get a head start by sowing some early seeds in the greenhouse indoors and by bringing them on so that they are ready to plant out as soon as the weather is mild enough.

EARLY WINTER

In the greenhouse

- Water vines and tree fruits throughout the winter

In the fruit garden

- Plant bare-root apples, apricots, sour cherries, pears, and raspberries
- Protect strawberries with hoop houses
- Prune apples, apricots, peaches and nectarines, pears, currants, blueberries, gooseberries, and grapes
- Raise new plants of raspberries
- Mulch red and white currants
- Force rhubarb by planting roots lifted in late autumn in containers of moist potting soil

In the vegetable garden

- Force endive
- Blanch endive
- Protect first-early peas and winter spinach with hoop houses and cloches
- Harvest Jerusalem artichokes, Brussels sprouts (and tops), fall and winter cabbage, savoy, red cabbage, parsnips, rutabagas, turnips, salsify and black salsify, perpetual spinach, endive, land cress, lettuce, winter radish, celery, celeriac, leeks, onions, winter spinach, and Swiss chard

MIDWINTER

In the vegetable garden

- Sow in the greenhouse or indoors seeds of cauliflower
- Harvest Jerusalem artichokes, Brussels sprouts, fall and winter cabbage, savoy, red cabbage, kale, parsnips, rutabagas, perpetual spinach, endive, mache, land cress, celery, celeriac, leeks, and winter spinach

LATE WINTER

In the greenhouse

- Change the border soil
- Sow seeds of eggplant and cucumber
- Prune vines and tree fruits

In the vegetable garden

- Sow indoors seeds of Brussels sprouts, summer cabbage, cauliflower, broccoli, broccoli raab, early varieties of carrot, turnips, lettuce, radish, and onions from seed
- Prepare seed potatoes for planting by presprouting in a cool place
- Harvest Jerusalem artichokes, Brussels sprouts, fall/winter cabbage, savoy, red cabbage, kale, parsnips, rutabagas, perpetual spinach, endive, mache, lettuce, celery, leeks, and winter spinach

GROWING VEGETABLES

Growing Vegetables

Cultivating your own vegetables is one of the best money-saving activities in the garden and is often the first step for anyone thinking of trying to become more self-sufficient.

With some careful planning, even a small homestead property can yield enough vegetables to supply a family of four or five year-round. Some vegetables lend themselves to quick and easy methods of storage (*see* pages 207–11), so that they can still be eaten fresh or in their natural state for most of the year. Others may be dried or salted, and nearly all will freeze successfully. Your own frozen produce will taste better than supermarket equivalents; not just because it represents your own hard work, but because it can be preserved within moments of being harvested (if you are organized), maintaining freshness and flavor.

Use common sense and instinct

Most vegetables are not difficult to grow; if you do find something particularly challenging, abandon it in favor of others that suit your methods, your schedule, and your garden better. Climate is key: You must choose the right crops for your conditions or you will do twice the work for far fewer results.

Follow the instructions given on the pages in this chapter but use your own common sense, and adapt as needed for your climate, soil, space, and habits. Few vegetables are so exacting in their growing requirements that they cannot accommodate small changes in timing or cultural conditions.

If you are a beginner, start small, with simple crops, like fast-growing, foolproof lettuce, then expand the range of crops you grow as you gain confidence and knowledge.

The weather is always a critical factor in any growing activity. It will often mean that sowing, planting out, transplanting, and later, harvesting, have to be delayed because conditions are not right or may be rushed if crops have matured faster than usual.

Do not rigorously follow the sowing, tending, and harvesting times given in this (or any other) book or on seed packs, but follow your instincts and be guided by experience as you grow in confidence. Far better to delay sowing, for example, for a week or two than to put the seed in wet, cold soil where it will not germinate.

In the same way, planting distances between rows and plants, or distances in which to thin out seedlings, are always approximate. It is a good idea to measure accurately when planning exactly what and how much to grow in your plot, but when doing the actual planting, most vegetable growers will judge spacings by eye, using experience to adapt them to suit the garden.

Pests and diseases

The pests and diseases outlined at the end of each group of vegetables are seldom as much of a problem as they appear. You would be unlucky to experience most of those listed. Few crops fail entirely because of disease or insect infestation, particularly if you tackle problems as soon as they arise. For more advice on diagnosing and treating plant problems, see the pests and diseases chapter on pages 140–51.

The main vegetable groups

Most individual vegetables can be categorized within a group, with other similar crops. In this book the groups are covered as follows:

- COLE CROPS (Brassicas)
  Broccoli, broccoli raab, Brussels sprouts, cabbages, cauliflowers, kale, and kohlrabi
- ROOT CROPS
  Beets, carrots, parsnips, salsify and black salsify, perpetual spinach, rutabagas, and turnips
- LEGUMES, BEANS, AND PEAS
- POTATOES
- SALAD GREENS
  Arugula, cress, endive, cucumbers, lettuce, and radishes
- OTHER OUTDOOR CROPS
  Jerusalem artichokes, artichokes, asparagus, celery, celeriac, garlic, leeks, squash, zucchini, pumpkins, gourds, okra, onions, scallions, spinach, and corn
- HEAT-LOVING CROPS
  Eggplants, peppers, and tomatoes

Raising Vegetable Seedlings

Nearly all the vegetables on the following pages can be raised from seed. You can do this outdoors in temporary seed beds or in the crop's intended growing position in the vegetable garden. Or sow indoors in pots or trays.

The method you choose will depend on a number of factors, including the prevailing weather conditions, the amount of space you have in the greenhouse, whether you have room for a dedicated temporary seed bed, the type of seed, and when you want to harvest each particular crop.

Asparagus, artichokes, and Jerusalem artichokes are usually raised from crowns, divisions, and tubers respectively, and potatoes are grown from seed potatoes. These should all be planted directly where they are to grow, as they cannot be transplanted successfully during their growing period.

SOWING UNDER COVER IN POTS OR TRAYS

Giving seeds a start in potting soil is a way of starting crops in cold climates, and is a good option for those vegetables that take a long time to mature, leaving their space in the vegetable plot vacant for a while for quicker-growing crops.

This method is also useful for crops that are better raised individually, and tender plants that need higher temperatures to germinate.

Seeds grown this way must be planted in special soil-based or soilless seed or potting mixes

SOWING OUTDOORS

Most vegetable seeds can be raised outdoors, although it is harder to produce early crops. Most seed packs will recommend a planting month or range of months in the instructions, but be guided by conditions where you live. Seeds need moisture and warmth to germinate, so the ground should be just damp (crumbly on the surface), not wet and cold.

The busiest sowing season is spring, but if you sow a little at a time through summer and into fall, your crops will mature in stages, rather than as a glut of produce all ready at the same time.

1 Break up large lumps of soil, rake it over, and remove any large stones. Add soil amendments and rake again to give a smooth surface.

2 Mark the correct planting distances along the bed and position a string line along the first row to give you a straight guide to work to.

3 Create a seed drill the required depth using a draw hoe. Pull it down the soil in a succession of strokes, rather than dragging it the entire length of the row in one pass.

4 If the ground shows any signs of being dry, water the bottom of the drill well. Put in any granular fertilizer you wish to use, according to the crop you are sowing.

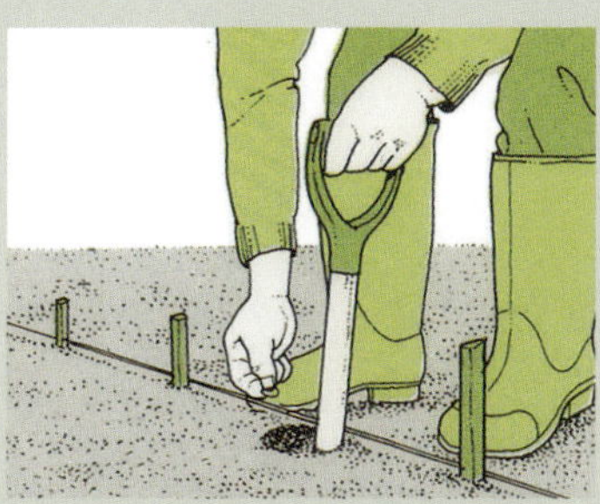

5 Using a dibber, sow the seed thinly. Even small seeds are best planted about two to each 1 in (2.5 cm). Use wooden stakes as distance guides.

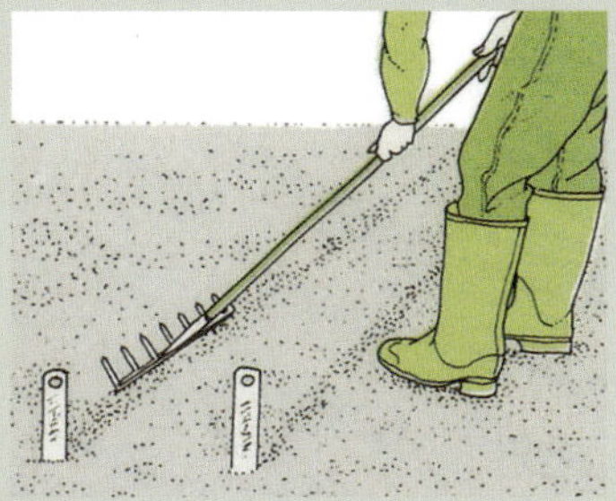

6 Pull the soil back over the seeds and firm it gently with the back of the rake. Put labels at both ends of the row to remind you what you have planted.

RAISING SEEDS IN A PROPAGATION UNIT

Heated propagation units maintain the temperature of the soil at the ideal level for germination and growth. But any covered container can be used to propagate seeds, from seed trays with a clear lid to a simple pot sealed inside a plastic bag.

Once seedlings are established, remove the lid and switch off the heat to an electric unit. Left to grow in the warm and humid conditions of a propagation unit, seedlings will become tall and straggly, with weak stem growth and they are likely to fail once you try to transplant them. With the lid removed, seedlings will grow at a slower rate, but will bush out and produce stronger young plants.

1 Squeeze some growing medium into a ball; it is wet enough to use if it cracks. For soilless mixes some moisture should come out.

2 Place the medium in a seed tray or propagation tray and firm it down with your hand or the back of the trowel to about 1 in (2.5 cm) from the top of the tray.

3 Sprinkle seeds over the medium, keeping them ½ in (1 cm) apart; position larger seeds individually. Cover with a thin layer of medium, sieved if it is soil-based.

4 Cover the seed tray with a lid, a pane of glass, or clear plastic bag, and place it in a sunny spot.

5 Remove the cover when the seedlings begin to show through the surface, and thin them out when they are large enough to handle.

available at all garden centers. These have been specially prepared to provide the seeds with exactly the right growing medium. Do not use garden soil—it contains too many weed seeds and is usually not a suitable texture for using in small containers. You can make your own seed and potting mixtures using loam (if you can get it), sterilized peat, and sand; but it generally makes more sense to buy it.

Plastic containers are typically better than those made of wood or terracotta because they retain water better and this will cut down on how much you need to water the seedlings. Whenever possible, use fairly shallow trays as they will use less potting mix—the seedlings will not need much root space before they are ready to be potted on or planted out. Other options include peat or coir pots, or even biodegradeable seed trays.

THINNING AND PRICKING OUT

When seedlings begin to appear, they will probably need to be thinned or pricked out to avoid overcrowding, which results in weak, straggly growth and a higher risk of disease.

With outdoor plantings, if you were careful with your sowing there should be little thinning required and the minimum of disturbance to the growing seedlings. You should always sow enough to allow for some seeds to fail, so some thinning out will be inevitable, and you should do this in two or three stages.

The first thinning should be as soon as the seedlings are large enough to handle, and they should be thinned to a distance of about

Pricking out seedlings will give them more room to grow into sturdy young plants. Hold them gently by a leaf and ease them out of the potting soil with a thin stick.

2 in (5 cm) apart. The second thinning will be a few weeks later to 6 in (15 cm) apart and the third to the final, recommended spacing.

Seedlings from the first thinning are best discarded onto the compost pile; those from a second thinning can sometimes be transplanted if space permits; and those from the final thinning may produce actual baby vegetables—tiny carrots and turnips, for example—that you can use in salads or cook lightly.

Thin seedlings when the soil is damp; this reduces the disturbance to those left behind. Pull up by their leaves, not the stem, if you intend to replant them—a crushed leaf will not harm the plant as much as a damaged stem.

Pricking out

Seedlings in trays must be pricked out when they start to crowd each other. Gently hold a leaf and loosen the roots with a thin stick or a pencil. Plant them into new containers by making a small hole and burying them up to the first set of leaves.

TRANSPLANTING INTO THE GROUND

When seedlings are large enough to move from their containers or a temporary seed bed to their permanent cropping location, it is time to transplant them. This should be done carefully in order to lessen the shock and avoid damaging the stem or the leaves too badly. Don't move plants directly from a warm greenhouse into the soil on a chilly day. Make sure each plant has a strong growing tip in the center before you transplant it—discard any that are unlikely to grow well.

1 Water the plants and new site well the day before. Choose a day that is mild and damp, but not windy. Dig up developing plants when they are about 4 in (10 cm) high, keeping some soil around their roots as you lift them out.

2 If the young plants are in containers, either dig them out of the seed tray or, if they are in individual pots, tip the pot upside down, supporting the plant carefully in your hand. If you use homemade pots from lengths of cardboard tube (above) or pots made out of newspaper, these can be put directly into the soil.

3 Make planting holes with a dibber or a hand trowel, ensuring they are slightly larger than the rootball with the soil attached and are set the correct distance apart.

4 Carefully set the roots in the hole and fill in the sides with soil. The plant should be set slightly deeper in the ground than it was before.

5 Firm in each plant and then gently tug a leaf to ensure that the plant is secure in the ground. Water well, with a light spray from the hose or a watering can to give a gentle flow.

Growing Cole Crops

Although only seven different vegetables fall under the heading of cabbage family crops—broccoli, broccoli raab, Brussels sprouts, cabbage, cauliflower, kale, and kohlrabi—the variety of shape, size, and taste they offer is enormous.

There are so many types of cabbage that in many areas it is possible to harvest them the whole year round and most are very easy to grow. Lightly cooked or grated raw, cabbages do not deserve their former reputation as an insipid, waterlogged vegetable. Broccoli comes in two basic types: broccoli and broccoli raab. The latter has a number of heads that are also called spears or florets.

Specialty catalogs may also sell seeds of purple sprouting broccoli, which has abundant purple-headed spears. It is also easy to grow and ready for picking in the late winter or early spring—a time when fresh produce from the garden can be in short supply. Its close relation, calabrese, yields large heads of tightly packed green flower buds in the late summer and early fall. It is similar to regular broccoli, but with a shorter growing season and is less hardy.

Cauliflowers are the most difficult of the cole crops to grow, but with care and experience can be harvested all year round. Different varieties are available, some with heads that are green, orange, or purple. Those with little pointed curds of pale green are known as romanesco cauliflower, although these are sometimes listed with broccoli in seed catalogs.

Brussels sprouts are among the most dependable of winter vegetables. Early spring sowings will yield fall crops, but if you really want to eat sprouts out of season it may be better to freeze some of the winter crop (which is easy) and use your precious land for a crop that is quicker to mature. In mild regions, Brussels sprouts can be harvested through the winter.

Kale is enjoying a resurgence in popularity. Its distinctive taste, which is even better after a frost, has much to offer and it can be harvested throughout the lean winter months. Different varieties have crinkly or smooth, but deeply serrated, leaves. Curly kale is an extremely hardy winter crop, which will withstand very cold conditions and will provide you with fresh vegetables during winter and early spring.

Kohlrabi is another unusual vegetable. The round purple or green growths, with their long, tough, leaf-bearing stems, rest on the surface of the ground, where they can remain until they are needed. They are best harvested when they are the size of a tennis ball, though, as they can turn fibrous if left too long. Better to freeze them for storage or trim away

GROWING TIMES AND HARVESTING TIMES

All cole crops are cool-season vegetables and in many regions they can be produced in your garden in both spring and fall or winter. The cropping season of others can be extended either by making sowings in succession over a period of weeks or months or by choosing to grow varieties that take varying times to mature (some have been specially produced to mature quickly).

Below is a chart which gives the shortest time each type of crop will take to reach full maturity and the months in which they can be harvested (subject, of course, to timed sowing).

VEGETABLE	SHORTEST GROWING TIME	HARVESTING TIMES
Broccoli	4 months	Late winter–spring
Broccoli raab	2 months	Late summer–fall
Brussels sprouts	6 months	Fall–late winter (best in winter)
Cabbage		
Early	4 months	Summer–fall
Late	At least 6 months	Fall–midwinter
Savoy	At least 5 months	Fall–spring
Spring	8 months	Spring–late spring
Red	8 months	Fall–late spring
Chinese	3 months	Fall
Cauliflower	At least 4 months (most varieties take longer)	Can be harvested all year round; easiest to produce for fall and winter
Kale	2–3 months	Winter–midspring
Kohlrabi	2 months	Late spring–fall

leaves and roots and keep them somewhere cool and dry. They taste like turnip and can be cooked in the same way or used raw in salads.

Choosing varieties

As well as the many different types of cole crops, each type has a wide choice of varieties, suited to different soil conditions, weather, and timing. The best guide before buying is to read the seed pack, or to check with other nearby gardeners and your local Cooperative Extension Office. Then experiment to see which suit both your garden and your palate best. Try early and late varieties, and consider the new hybrids that appear on the market each year.

CULTIVATION

Comprehensive instructions for sowing, growing, and harvesting cole crops are given in the chart overleaf. All cabbage family crops like slightly alkaline, rather than acidic, soil, so unless your soil is alkaline, amend it with lime before planting to improve conditions.

Most cole crops thrive in good fertile soil, but often do best if they follow another crop for which the soil was enriched. As most types take a number of months to mature and become ready for harvesting, it is a good idea to sow them in temporary beds, leaving room in the main vegetable beds for some quicker-growing crops that can be harvested before the cole crops are ready to transplant.

After harvesting, plants should be dug up and discarded. The leaves and stems can be composted, but cut up or crush anything woody first, as the thick stems can take years to break down.

CULTIVATION OF COLE CROPS

3

1 Because most cole crops are slow-growing, you can sow the seeds in a temporary bed first.

2 Thin seedlings as they appear. Transplant carefully when large enough to handle.

3 Plant out large crops in their final cropping site on a mild, damp day, using a string as a guide.

4 Support tall plants, such as Brussels sprouts (below) and broccoli, with large stakes, tied in with twine.

4

5 Harvest crops as soon as they are ready, when cabbage hearts have firmed and before sprouts start to open out.

6 Cabbages growing in the winter may need to be protected from harsh weather with cloches. All cole crops may need netting to protect them from birds.

6

PESTS AND DISEASES

Cole crops are susceptible to a number of pests and diseases. Many of the diseases can be controlled or eliminated by treating the soil with commercial products such as horticultural oil or insecticidal soap.

Cole crops, broccoli in particular, are irresistible to hungry birds. Protect young seedlings by stretching black cotton along the rows. It may even be necessary to net broccoli plants when they are maturing.

Slugs and snails can often be a problem too, eating their way quickly through an entire line of seedlings. They can be controlled by scattering slug pellets on the ground. However, these pellets contain poison that can be harmful to animals, so put them under a pot and prop up one side of it, so only the slugs can enter. Better still, try to encourage animals into

CULTIVATION OF COLE CROPS

VEGETABLE	SOIL/POSITION	SOWING	DRILL DEPTH
Purple sprouting broccoli and calabrese	**Soil:** Firm, heavy, loamy soil that is well-dug and enriched **Position:** Open and sunny, but with some shelter	Sow in temporary beds late spring. Calabrese sowings can begin in late winter	½ in (1 cm)
Brussels sprouts	**Soil:** Rich loam—a firm soil that has been enriched with well-rotted compost or manure **Position:** Prefer an open site, but do not mind partial shade	Sow indoors in late winter for early crops. In temporary beds from midwinter to spring	½ in (1 cm)
Cabbage	**Soil:** All types like well-drained, fertile, non-acid soil, preferably following a crop for which the ground was enriched. They will grow successfully in most well-cultivated soils **Position:** Open and sunny	**Early cabbage:** Sow indoors or in cold frames in late winter. In temporary beds from late spring. In warm regions, sow in late summer for a winter crop **Late cabbage:** Sow in temporary beds in spring or in fall for winter crops **Savoy:** See late cabbage **Chinese cabbage:** Sow in permanent site in spring or fall	½ in (1 cm)
Cauliflower and broccoli	**Soil:** Deep, well-drained, well-cultivated, rich soil **Position:** Open and sunny, but sheltered	**Early varieties** **Cauliflower:** Indoors in spring or in midsummer for late fall and winter harvest **Broccoli:** Sow indoors in late winter for transplanting in spring. For fall crops, sow directly in the garden in mid- to late summer	¾ in (2 cm)
Kale	**Soil:** Rich loam but will grow in most well-cultivated soil **Position:** Any, but particularly likes an exposed site	Sow outdoors in cropping site in fall for winter harvest in mild-winter regions. Elsewhere, sow in early- to midsummer for winter harvest	½ in (1 cm)
Kohlrabi	**Soil:** Fertile, well-drained **Position:** Any, but not too shaded	Sow outdoors in cropping site at monthly intervals from early spring	½ in (1 cm)

your garden that will control your pests for you, such as chickens.

Good garden hygiene will do much to keep disease at bay. This means ensuring that dead plants are removed from the ground and composted at once. Never leave them in the soil once they have yielded their full harvest. Equally, discard all infected or diseased plants and ensure all the tools and equipment you use are clean.

Some of the common pests and diseases that attack the cabbage family are listed here; for more information on identification and treatment, see the relevant pages: cabbage root maggot (page 148), cabbage caterpillars (page 144), flea beetle (page 144), cabbage whitefly (page 150), clubroot (page 148), downy mildew (page 146), wirestem (page 148), and whiptail (page 145).

PLANTING OUT/TRANSPLANTING	PLANTING DISTANCES (between plants/rows)	CULTIVATION	HARVESTING
Thin regularly; plant in cropping site when plants are 4–6 in (10–15 cm) tall (about six weeks after sowing)	**Broccoli:** 2 ft x 2 ft 6 in (60 x 65 cm) **Calabrese:** 18 in x 2 ft (45 x 60 cm) Calabrese plants can be planted closer together to give a large harvest of small spears	Hoe regularly between rows to eliminate weeds. Protect from birds by netting. Water well if at all dry. Stake plants, tie loosely to support if weather is windy, and hill up soil around the stems for extra support	In both cases, cut while flower buds are still tightly closed. Cut from the center, taking central heads with 4–5 in (10–12.5 cm) of stem. Cut just above a side shoot to encourage more growth. Harvest sideshoots as they appear
Thin regularly; plant in cropping site when plants are 4–6 in (10–15 cm) tall (about six weeks after sowing)	2 ft 6 in x 2 ft 6 in (65 x 65 cm). Stagger the rows	Hill up soil around stems a month or so after planting to firm plants. Stake plants and tie loosely. Water freely. Break off the lower leaves as they turn yellow	Pick while still small and tightly knotted, preferably after the first frost. Start at the bottom of the stem. Pick off top leaves and cook as greens (this encourages sprouts at the top of the plant)
Early cabbage: Those sown indoors should be planted out spring **Late and savoy cabbage:** Should be transplanted to cropping site when 4–6 in (10–15 cm) tall (about six weeks after sowing) **Chinese cabbage:** Thin seedlings progressively, so they are correct distance apart when plants are 4 in (10 cm) tall	**Early cabbage:** 18 x 18 in (45 x 45 cm) **Late and savoy cabbage:** 2 x 2 ft (60 x 60 cm) **Chinese cabbage:** 1 ft x 1 ft 6 in (30 x 45 cm)	**For all cabbages:** Hoe regularly between rows to discourage weeds and keep plants well-watered. Remove any decaying leaves as they appear **Chinese cabbage:** Gently pull any drooping outside leaves up around inner ones and tie loosely to encourage thicker growth inside	**For all cabbages:** Cut when heads are firm and leave stumps in ground (except early ones) to give a few extra subsequent cuts of greens **Early cabbage:** Cut cabbages in spring and use as tender greens
Early varieties cauliflower: Prick out those grown indoors and put in cold frame to harden off Plant all varieties in cropping site when 4–6 in (10–15 cm), using a dibber	**Early varieties**: 18 x 18 in (45 x 45 cm) **Winter varieties:** 2 ft x 2 ft 6 in (60 x 75 cm) in staggered rows	Hoe between rows to discourage weeds. Water frequently if at all dry. When curds form, bend outside leaves over them. This protects from sun, frost, or staining from soil—all of which discolor the curds	Cut when curds are formed, but are tightly packed. Cauliflowers are best picked in the morning, when the dew is still on them. If more ripen at one time than you need, pull them up and hang by the roots in a cool, dry place. They will keep for up to three weeks
Thin as seedlings appear to recommended distances	2 x 2 ft (60 x 60 cm)	Hoe between rows and plants to eliminate weeds. Tread round plants to firm ground. Stake and tie if necessary	Harvest from the plants' center to encourage more growth; frost can improve the taste
Thin progressively until required planting distance is achieved	9 x 15 in (25 x 40 cm)	Weed between rows and keep plants well-watered	Pull out bulbs (which you can see on the ground) when about 3 in (7.5 cm) diameter. If larger than this, they will be woody

Growing Root Vegetables

The term "root crops" covers a wealth of vegetables. Included here are beets, carrots, parsnips, salsify and black salsify, perpetual spinach, Swiss chard, rutabagas, turnips, and sweet potatoes.

Rutabagas and turnips are, strictly speaking, members of the cabbage family, but their growing habits and requirements mean that they are often grouped with root vegetables.

Sweet potatoes are also not true root crops; and, despite their name, they are only very distantly related to the common potato, with quite different cultivation requirements, so they are rarely grown together. They are actually a herbaceous perennial vine, although they are grown from tubers or "slips," pieces of the tuber that will produce roots and new plants.

Beets are most frequently thought of as a salad vegetable, cooked and then eaten cold, or pickled, but they are also delicious eaten hot or in soup and can be grated and eaten raw (the young ones are the most palatable). There are many different-shaped varieties, and also some that are golden in color.

Carrots come in myriad shapes from round, or finger-shaped, to conical, or elongated—far more interesting than the supermarket variety. Like most roots, they store well (*see* page 207), so can generally be eaten fresh all the year round. Their cooking possibilities are endless, even as a sweetener in some deserts and they make a deliciously moist cake.

Parsnips, too, have more uses than just as a side vegetable. They are one of the most common ingredients of homemade wine. Varieties bred to yield small roots are very useful for people who have only a small amount of land to devote to vegetable production.

Try something different

Salsify and black salsify are two unusual root vegetables that are worth growing for the tasty interest they can give to family meals. Both produce long, thin roots; salsify looks a little like an elongated parsnip, while black salsify is similar in shape but has a black skin. They take up quite a lot of land for a good six months or more, but if you have space for them in your overall vegetable garden plan, they make an interesting change on the table.

Perpetual spinach is not as well-known as it deserves to be. It is sometimes known as spinach beet and is the easiest of all spinaches to grow, having a higher yield and a greater tolerance to dry conditions. It also has a less acidic taste than spinach. As its name suggests, perpetual spinach can be harvested for most of the year.

Swiss chard is better known. It is similar to perpetual spinach, with wider leaf stalks and red, white, and golden-ribbed varieties are available. The green leafy part may be torn from the stem to be cooked and eaten as greens, and the midrib can be cooked and served on its own.

Beets are a rich source of vitamin B, when eaten raw—grate them into salads to make the most of their goodness. The roots store well in a cool place or can be pickled.

GROWING TIMES AND HARVESTING MONTHS

Continuous crops of roots can be produced in the garden by regular sowings and experimenting with the various early- and late-maturing varieties. Many root crops are sown in two main stages: one planned to give the early, small, sweet crops, with a heavier sowing to follow, giving the main crop through the winter.

VEGETABLE	SHORTEST GROWING TIME	HARVESTING MONTHS
Beets	4 months	Early summer–fall
Carrot	3–4 months	Early summer–fall
Parsnip	7–8 months	Fall–winter
Salsify and black salsify	6–7 months	Fall–early winter
Swiss chard	3–4 months	Late summer–winter
Perpetual spinach	4 months	Late summer–spring
Rutabaga	Early sowings need 3 months. Later spring/ summer sowings take about 5 months	Late summer–spring
Sweet potato	4–5 months	Fall
Turnip	At least 2 months	Summer–early winter (Greens: Winter–spring)

Earthy crops for warming stews

The rutabaga has a surprisingly delicate taste for such a hardy vegetable. If you have more space in the ground than in your storage sheds, the crop can be left in the ground all winter with no ill effects.

Turnips come in all kinds of shapes, sizes, and colors. They may be round like a ball or flattened on the top and bottom. In color, they may be white or golden, or green, red, or purple-topped. They grow so fast that they make an excellent "intercrop" between rows of longer maturing vegetables and they can also be grown for their leaves, harvested through the winter and early spring, which can be cooked like spring greens.

Sweet potatoes will not reliably grow outdoors in colder regions, although new varieties are being bred all the time that make them worth a try. But they will grow in containers, in beds in the greenhouse, or outdoors in milder climates. Do not try to grow plants from tubers bought in the supermarket, as they are likely to be tender varieties from overseas; instead, look for tubers or slips sold by mail order. They may have pinkish or orange flesh and produce attractive flowers, similar to the morning glory vine.

CULTIVATION TIPS

The cultivation details for the root crops included in this section are given on pages 70–71. Most prefer a similar type of soil and not one that has been freshly enriched, which encourages the roots to fork and split. Rocky ground can also cause forking of the roots as they grow around obstacles in the soil.

Hoe carefully between rows. Keeping weeds at bay is important, but it is easy to damage the tops of the developing roots, which are usually very close to the surface.

Sow seeds of root crops straight into their permanent positions. Make sure the soil is well prepared for sowing and is dry and crumbly; it is better to delay planting than to sow seeds into wet, sticky, cold soil.

Getting a double yield

Most root crops take less time to mature than cole crops; even so, an intercrop of quicker-maturing vegetables such as lettuce can be sown between the rows of the longer-maturing varieties in the early stages of growth. Another way of making your garden more productive is to sow radish seed together with

CULTIVATING ROOT CROPS

1 Prepare the soil by raking in an all-purpose fertilizer. Sow the seed in shallow drills, as instructed on the packet. Leave the correct spacing between rows.

2 Water young plants well to prevent the developing roots from forking in search of water.

3 Regularly weed between rows of growing plants, taking care not to damage the growing roots.

4 Thin plants progressively. The third thinning may produce tiny vegetables.

parsnip along the same rows. The radish seeds germinate much faster than the parsnips, so they indicate where the parsnip seedlings will later emerge, as well as giving an extra crop of radishes that you can harvest before the parsnips start to crowd them out.

Routine care

All root crops should be watered freely if there is any danger of the soil drying out. Those deprived of water will be woody and fibrous as a result; they may also fork if allowed to dry out and are then subjected to a heavy rainfall. Keep irrigation consistent and regular. Drip irrigation is an ideal choice for root crops, as it avoids the swings between too much and too little water.

Hoeing between the rows is necessary to keep down the weed population, but take extra care not to chop through the developing vegetables with the hoe. If necessary, when your root crops are young, you may need to weed by hand.

Harvesting and storage

Most roots can be harvested over a fairly long period, the early pullings providing small, but delicious, sweet-tasting baby vegetables. Pull these by hand and then lift later crops with a fork, being very careful not to damage and bruise the roots.

Some, such as parsnip and rutabaga can be left in the ground until they are wanted (provided you do not want the bed for anything else) and their flavor is actually often improved by enduring a frost. Others, such as turnips, should not be left in the ground and should be lifted and stored (*see* page 207). All root vegetables are among the best types of crops for storing, although they also freeze well.

Sow intercrops between slow-maturing root vegetables. Here, lettuces are growing beside the carrots to their left, but will be harvested before the carrot foliage gets bushy.

PESTS AND DISEASES

Root crops are prone to pests and diseases, which can often be controlled with good garden management. Consult your local garden center or Cooperative Extension Office for advice on tackling a particular problem.

Paying attention to good garden hygiene and destroying infected plants immediately (*see* page 143) will help to keep your crop healthy and strong enough to shrug off disease and pest attacks.

One of the most virulent pests is the carrot rust fly (which, despite its name, may also attack parsnips). Early crops of carrots are likely to escape attack as they may be harvested before the maggots hatch and burrow into the ground in search of the tasty roots. The carrot fly is attracted by the smell given off by leaves when handled, so avoid touching the plants as much as possible. For the same reason, destroy any thinnings, or remove them from the ground immediately. Cover crops with row covers or build a 30 in (75 cm) enclosure around your carrots to prevent the flies from laying their eggs on the crop.

Flea beetles present a danger to rutabagas and turnips. Avoid growing all these crops on soil which you know is harboring clubroot, because the vegetables will be swollen, distorted, and inedible. Leaf miners and aphids may attack beets.

Some of the common pests and diseases that attack root crops are listed here; for more information on identification and treatment, see the relevant pages: carrot fly (page 151), beet fly (page 151), wireworms (page 151), black slugs (page 151), celery fly (page 145), cutworm (page 147), rust (page 146), white blister (page 145), clubroot (page 148), and violet root rot (page 148).

CULTIVATION OF ROOT CROPS

VEGETABLE	SOIL/POSITION	SOWING	DRILL DEPTH
Beets	**Soil:** Light, but deep loam that has been well-dug and cultivated **Position:** Will grow in most places, but avoid heavily shaded sites	**Globe varieties:** Sow indoors in spring. Harden off before transplanting. Sow in permanent site in spring and fall **Long-rooted varieties:** Sow in regular succession in cropping site in early summer. Beets grows from clusters of three or four seeds. Space these about 2 in (5 cm) apart	½ in (1 cm) ¾–1 in (2–2.5 cm)
Carrot	**Soil:** Well-cultivated and deeply dug sandy loam, but will grow in any light soil, providing it is free of any large lumps of dirt or big stones **Position:** Any that is not heavily shaded. Early crops like an open, sunny spot	**Fingerlings:** Sow in the ground in early spring **Maincrop:** Sow in regular succession outdoors from midspring through summer. Sow two seeds at a time, 1 in (2.5 cm) between pairs	¼ in (0.5 cm)
Parsnip	**Soil:** Deeply dug, light loam, free from large stones **Position:** Any which is not heavily shaded	Sow outdoors in early spring. The surface of the soil must be dry and fine	1 in (2.5 cm)
Salsify and black salsify	**Soil:** Deeply dug, light loam **Position:** Open and sunny	Sow outdoors from midspring to late spring	¾ in (2 cm)
Swiss chard and perpetual spinach	**Soil:** Any well-cultivated soil; these vegetables particularly dislike poor drainage **Position:** Any that is not heavily shaded	**Perpetual spinach:** Sow outdoors in spring and again in late summer **Swiss chard:** Sow outdoors late spring through summer	1 in (2.5 cm)
Rutabaga	**Soil:** Light, fertile loam that is well-drained but not too dry. Does not like acid soil **Position:** Open and sunny. React badly to overshadowing	For an early crop, sow outdoors in early spring; otherwise sow outdoors in early summer. Sow two seeds to 1 in (2.5 cm)	¾ in (2 cm)
Turnip	**Soil:** Well-cultivated, light loam **Position:** *see* Rutabaga	Sow indoors in late winter. Then outdoors in regular succession from midspring to late spring through summer. Late sowings provide fall crops **Turnips for greens:** Sow outdoors at end of summer	½ in (1 cm)
Sweet potato	**Soil:** Fertile and free-draining, with well-rotted manure or compost dug in the previous fall. Can be planted on hills or mounds of soil **Position:** Sheltered spot in full sun	Pot cuttings in mid-spring/early summer. Keep slips under glass or in a propagator until roots appear, then pot into individual pots of potting soil	½ in (1 cm)

THINNING/TRANSPLANTING	PLANTING DISTANCES (between plants/rows)	CULTIVATION	HARVESTING
Thin seedlings indoors to 6 in (15 cm). Plant out late spring. Thin seedlings when about 1 in (2.5 cm) high then again to final spacing when roots are 1 in (2.5 cm) in diameter.	**Globe:** 5 x 12 in (12.5 x 30 cm) **Long-rooted:** 8 x 18 in (20 x 45 cm)	Protect seedlings from birds and slugs. As they grow, hoe between rows and water plants well	Begin by pulling out every other plant. Those that remain can be left until wanted, but globe varieties should not grow more than 3 in (7.5 cm) across or they will turn woody. Lift in late fall before the frosts. Twist off the leaves just above the root to seal in the juices
Fingerlings: Thin to 1 in (2.5 cm) apart, then 3 in (7.5 cm). **Maincrop:** Thin to 3 in (7.5 cm) when first true leaves appear, then to final spacing when carrots are as thick as a finger. Never leave thinnings on the ground—they attract carrot rust fly	**Fingerlings**: 3 x 9 in (7.5 x 22 cm) **Maincrop:** 6 x 12 in (15 x 30 cm)	Water after sowing to aid germination. After thinning hill up soil around base (top of carrot should not show through soil). Hoe between rows to check weeds	**Fingerlings:** Pull every other plant. Pull young carrots by hand **Maincrop:** Lift carefully with a fork, no later than midfall. If storing (*see* page 207), trim off the top (compost this immediately), remove the soil from the carrots, but do not wash them before they are stored, or they will rot
Thin to 2 in (5 cm), then to final recommended spacing. Do not try to transplant thinnings as it is rarely successful	6 x 15 in (15 x 40 cm)	Hoe between the rows to keep weeds in check. Water plants if they are at all dry	Dig up parsnips as wanted—the flavor improves after the first frost. They can stay in the garden until spring, otherwise, dig them up and store them
Thin twice to recommended distance	**Salsify:** 9 x 12 in (20 x 30 cm) **Black salsify:** 12 x 12 in (30 x 30 cm)	Hoe between rows and water plants in dry spells to prevent them going to seed. Pull the soil up around the base of the plants	Lift carefully from midfall onwards. Although both are hardy and can be left in the ground, it is better to harvest them and freeze them. If space permits, you can leave some salsify in the ground until spring; the new growth can be picked and cooked like spring greens
Thin progressively to recommended distance, beginning as soon as the plants are large enough to handle. Later thinnings can be eaten	6 x 9 in (15 x 20 cm)	Hoe to eliminate weeds. Water plants generously if there is any danger of them drying out	Pick the early sowing in summer and the later sowing through fall and winter. Do not pick fall plants too heavily if you want a good supply through the winter months. To harvest, take the outer leaves gently away from the root, leaving the young, central leaves to encourage more growth
Thin regularly to recommended distance	12 x 18 in (30 x 45 cm)	Hoe in the early stages to control weeds; this will not be necessary when the leaves cover the ground and smother the weeds	For small, sweet-tasting rutabagas, you can begin harvesting two to three months after the first sowing. Most will be allowed to grow bigger and can be harvested as needed. Leave in the ground through winter, or lift and store
Seedlings indoors: Thin when about 4 in (10 cm) high **Other sowings:** Thin to 4 in (10 cm) apart when first leaves appear. Thin to 8 in (20 cm) when turnips are large enough to make cooking worthwhile **Turnips for greens:** No need to thin, just leave plants to grow	8 x 12 in (20 x 30 cm) **Greens:** rows 9 in (20 cm) apart (*see* Transplanting)	Hoe to eliminate weeds	Pull early sowings when turnips are 2–3 in (5–7.5 cm) in diameter. Check by pulling the foliage aside to see the top of the turnip. Lift others in midfall to late fall. Twist off tops and store **Greens:** Cut leaves when they are about 8 in (20 cm) long. They can be picked in late fall and early winter, but are more often left until spring
Grow young plants in a frost-free place until midsummer. Harden off for 2–3 weeks before planting outside. Shelter with cloches or row covers for a further 2–3 weeks	12 x 30 in (30 x 75 cm)	Keep the plants well-watered, but not too much, as this can cause the tubers to split. Allow the foliage to spread over the soil to smother the weeds	Pick leaves and cook like spinach once at a usable height. Harvest tubers from end summer; they will rot in cold, wet soil. Cut off the vines before carefully lifting the tubers, which bruise easily. Do not wash until you want to cook them. The flavor improves if stored in a humid place for two weeks after lifting

Growing Beans and Peas

Although this section includes only two vegetables (also known as pulses or legumes), beans and peas are two of the most useful and varied crops in the vegetable garden. Both groups of plants are quick to mature and their roots manufacture valuable nitrogen, which they release into the soil, enriching it for the following year.

Peas and beans need stakes to support them as they grow. When mature, the plants will have completely smothered this framework with lush foliage and a crop of beans.

Purple-podded beans look stunning, although the beans inside are a conventional white and the pods usually lose their vibrant color when they are cooked.

Pulses yield their main harvest through the summer and fall, but both beans and peas can be grown especially for drying and all types also freeze well, so they can be enjoyed all year round.

GROWING TIMES AND HARVESTING MONTHS

Peas and beans are largely a summer and fall crop, but there is no reason why all types should not yield heavy harvests throughout these seasons. Many different varieties are available—both early- and late-maturing. Sow a good selection for maximum cropping, but make sure you sow the right variety at the correct time.

VEGETABLE	SHORTEST GROWING TIME	HARVESTING MONTHS
Broad beans (lima beans)	4 months	Late spring–summer
Bush and pole beans	2–3 months	Early summer–fall
Green beans	3 months	Late summer–fall
Soybeans	3½ months	Late summer
Peas		
Early:	2½ months	Early summer and fall
Second:	3½ months	Midsummer
Maincrop:	3 months	Midsummer–late summer
Late:	3 months	Late summer–fall
Snow peas	2½ months	Early summer–fall
Asparagus beans	3 months	Early summer–late summer

Choosing beans

Among the bean group are broad, bush, pole, and green beans; other beans are grown for drying. The large, kidney-shaped broad beans sometimes have a reputation for being tough, but eaten straight from the garden when they are young, they are delicious and mouth-wateringly tender. The top foliage can also be picked and cooked like spinach.

There are many varieties of broad beans, suitable for sowing at different times of the year. They also include dwarf varieties suitable for small vegetable gardens. Most types are hardy and will yield heavy crops.

Pole and bush beans are similar, although bush beans are low-growing, a little hardier, and produce very heavy yields. Both these beans are available in many varieties, from those with a waxy finish and yellow pods to deep purple varieties and those that are flat or round-podded. Other beans, such as lima and pinto, are grown specifically for drying and the white-seeded varieties are best for this. Borlotti, pea beans, cannellini beans, and many other beans are available through specialty suppliers or catalogs. Give dry-bean crops the longest possible growing season for the best yield.

Green beans usually mature after pole and bush beans, although they can be grown to be harvested simultaneously. They are easy to grow and pretty enough to be incorporated into a flower garden on a teepee system (*see* page 78). Dwarf varieties will also need stakes.

Soybeans are an excellent source of protein, particularly valuable to vegetarians, but do best in a warm site. Young pods can be eaten whole, but to pod more mature beans you should blanch them in boiling water first for five minutes, cool, then break open the pods and remove the beans inside before adding them to stir-fries and salads.

Pea varieties

Peas also produce endless varieties and different types will grow to different heights. They are divided into two main groups. Garden, or shelling, peas are grown for the seeds, which are removed from the pod before eating. Snap peas have thick pods that are eaten whole, seeds and all. Many varieties have different qualities. Some are hardier than others and can withstand earlier sowing, but a succession of planting can produce harvests of succulent young peas right through the summer and into fall. In mild regions peas can be grown well into the winter months. Wrinkle-seeded varieties are considered by some to be more flavorful, but they are less hardy than the round-seeded types.

In addition to the peas mentioned above, snow peas and petit pois are also very popular. Snow peas are harvested while the peas are barely formed and are eaten, pods and all. Petit pois are very small, sweet-tasting peas, which can be cooked in the pod and shelled afterwards, or shelled first like other peas. Less well-known is the asparagus bean, which is a member of the pea family, but looks quite different. It is sometimes known as the yardlong bean because of its long, crinkly, pods. Asparagus beans are cooked whole and get their name because they taste a little like asparagus.

CULTIVATION

All beans and peas grow best in soil that has been enriched for a previous crop. Alternatively, for peas and green beans in particular, dig some well-rotted manure into the ground a month or two before sowing. The roots will quickly grow down in search of this, giving the plants a firm anchorage.

Beans and peas are sown deeper into the soil than most other crops, the drills for sowing being a good 2 in (5 cm) deep. When sowing, the soil should be forked over to a depth of at least 4 in (10 cm), making sure that it is crumbly, with no big lumps or stones. Scrape out wide drills of 6–8 in (15–20 cm), using the width of the draw hoe blade, and rake the soil loosely back over the seeds.

Mice, rabbits, and squirrels often dig up and eat the seeds, but you can discourage them by sprinkling some holly leaves in the drill, setting traps along the rows, or keeping a good, hungry, farm cat.

Sow seeds at the correct distance apart, as no thinning is usually required, but to ensure a full crop, either sow two seeds at each station and pull out the weakest one as the seedlings emerge, or sow some extra seeds at the ends of the rows and use the resulting seedlings to fill in any gaps as the others germinate.

Snow peas are ready to be picked when the peas inside the pods are only just developing and they are eaten whole.

Thinning out is only necessary with beans and peas if you planted extra seeds to insure against the failure of a few.

Most beans and peas need some sort of support. Bush varieties can be left to grow straggly with no supports, but not only will the pods become very mud-splashed, they are not always easy to find in the tangle of foliage. Various ways of supporting them are shown on page 78.

Except for varieties grown specifically for drying, all peas and beans should be harvested when young for the best flavor. The more you pick a plant, the more pods it will produce, so pick heavily and regularly, freezing any surplus vegetables you may have.

Legume roots make valuable nitrogen, so they can be left in the soil after the plant has yielded its full harvest. Cut off stems at ground level and compost the leaves.

CULTIVATING BEANS AND PEAS

1 Sow seeds directly in evenly dug rows. Make sure the ground is watered and weed-free.

2 Protect the emerging seedlings against birds by covering them with chicken wire or cloches.

3 If you planted two or three seeds at each station, you will have to thin the seedlings.

4 Most peas and beans will need to be supported as they grow to stop them straggling. Train them up poles.

5 Pinch out the tips of broad bean plants to discourage black aphids. Destroy any infected leaves.

6 Most peas and beans are best when picked young and tender. This will also encourage the plants to grow.

PESTS AND DISEASES

Rabbits are one of the biggest threats to peas and beans. Slugs (page 147) will attack seedlings. If the plants are not well-watered, the flowers fail to set, pods do not form, and plants will wilt and die quickly. In wet weather, particularly if it is cold too, peas are vulnerable to mildew (*see* page 146).

The most common pests and diseases to affect beans and peas are listed here; turn to the relevant pages in the pests and diseases chapter for more information on diagnosis and management, red spider mite (page 146), thrips (page 150), blackfly (page 144), anthracnose (page 150), Mexican bean beetle (page 150), cucumber beetles (page 150), and weevils (page 150).

CULTIVATING BEANS AND PEAS

VEGETABLE	SOIL/POSITION	SOWING	DRILL DEPTH
Broad beans	**Soil:** Good, rich loam, but will grow in any well-cultivated, well-drained soil **Position:** Prefer a well-sheltered site	Sow outdoors at the recommended planting distances in fall for longpod varieties, late winter through spring for summer crops, and late spring for fall crops. In cold, exposed gardens sow indoors in boxes 2 in (5 cm) apart or singly in biodegradable pots in winter	2 in (5 cm)
Pole, bush, and drying beans	**Soil:** Light, well-drained, non-acid loam **Position:** Open and sunny, but sheltered	Sow indoors midspring to late spring: outdoors in May, or slightly earlier if protecting with cloches, then regularly until early summer. Soil must be 60°F (15°C) for beans to germinate. Put seeds in pairs 1 in (2.5 cm) apart or sow singly 3 in (7.5 cm) apart	1½–2 in (4–5 cm)
Green beans	**Soil:** Rich, well-cultivated, well-drained, deep soil **Position:** Open and sunny, but sheltered	Outdoors in late spring if well-protected with cloches or row covers. To be safe, plant in early May (mid to late May in cold regions) or indoors in April. Seed will not survive frosts	2 in (5 cm)
Soybeans	**Soil:** Well-cultivated, well-drained loam **Position:** Open and sunny; preferably warm	Sow outdoors in late spring, or wait until early summer in cooler regions. Sow seeds 3 in (8 cm) apart. Indoors, sow four seeds to a pot in early spring	1–2 in (2.5–5 cm)
Snap and shelling peas	**Soil:** Moist, medium loam that is well-cultivated, well-drained, and non-acid **Position:** Open and sunny	Flood drills with water if dry and sow when water has soaked through. Sow seeds 2–3 in (5–7.5 cm) apart or in a V-shaped drill, 3 in (7.5 cm) deep **First:** Sow outdoors in late fall (protect with cloches through winter) or late winter and again in early summer **Second:** In cropping site in spring **Maincrop and late:** Sow outdoors in late spring	2 in (5 cm)
Snow peas and petit pois	**Soil:** *see* Peas **Position:** *see* Peas	Outdoors in late spring through summer. Follow general instructions for sowing	2 in (5 cm)
Asparagus beans	**Soil:** *see* other beans, but soil must be really well-drained. **Position:** *see* beans	Outdoors late spring through summer. Sow two seeds together every 10 in (25 cm)	2 in (5 cm)

THINNING/TRANSPLANTING	PLANTING DISTANCES (between plants/rows)	CULTIVATION	HARVESTING
Outdoor sowings don't need thinning; fill any gaps with extra plants. Harden off indoor sowings from the end of March and plant in cropping site as soon as soil conditions allow, usually in April. Stand biodegradable pots in water first and plant so that the rim is beneath the soil surface	**Tall varieties:** Double rows 8 in (20 cm) apart with 8 in (20 cm) between plants. Stagger seeds in the two rows. Leave 2 ft (60 cm) between double rows **Dwarf varieties:** 9 in (20 cm) between plants; 18 in (45 cm) between rows	Protect seedlings from birds, stake growing plants, and hoe between rows to control weeds. Pinch out growing tips when plants are in full flower, when the first pods form, or when you see aphids (*see* page 144). This helps to control aphids and keep the plant bushy	Begin as soon as beans are big enough to cook or sooner, before the beans are properly formed, if you like to eat the pods as well. Pick the bottom clusters first and continue regularly picking as beans become ready. Old beans will be tough, but can be used for next year's seed
Harden off seedlings in boxes or under glass in early May and plant out at the end of the month. If sown outdoors, pull out the weaker seedlings as soon as germination has occurred, or thin to recommended distance	**Bush:** 6 x 15 in (15 x 40 cm) **Pole**: 9 x 18 in (20 x 45 cm)	Hoe to eliminate weeds and pile the soil up around the stem of the plants to firm them. Water plants freely and mulch with compost or well-rotted manure if ground is prone to drying out. Support larger plants	Begin harvesting when beans are 4 in (10 cm) long and pick regularly to encourage growth **Beans grown for drying:** Leave to ripen fully on plant. In late summer, pull up entire plant and hang upside down in a dry, airy place to finish drying. Shell beans and store them when fully dried
Indoor sowings: Harden off in late spring. Plant out in early summer **Outdoor sowings:** Those sown in cropping site need no thinning; fill any gaps with extra plants	Depends on method of support (*see* page 78). **For crossed poles:** Two staggered rows, plants 10 in (25 cm) apart; rows 2 ft 6 in (75 cm) apart. If more than two rows, allow 5 ft (1.5 m) between them **Vertical stakes:** Single rows 1 ft x 2 ft 6 in (30 x 75 cm) **Teepee:** Six plants spaced round 3 ft (1 m) circle	Decide how you will stake before you sow (page 78); supports should be 8–10 ft (2.5–3 m) high. Encourage plants to grow up them by twining emerging tendrils around stakes. Hoe between rows and water freely, mulching ground if it tends to dry. Spray foliage to help to set flowers. Pinch out top growing tips to encourage side growth	Pick regularly while pods are young and tender—freeze or pickle the surplus rather than leaving them to get tough on the plant. As soon as beans swell in the pod, they will make tough eating
No need for thinning. Plant out seedlings grown under glass in early summer	Single rows, 3 in (8 cm) between plants and 9–12 in (25–30 cm) between rows	Protect seedlings from birds with black cotton, netting, or hoop houses. Support plants with stakes	**For eating fresh:** Pick in late summer and fall when the pods are green. Harvest each plant all at once. When pods change to a creamy color, only eat the beans inside **For drying:** When foliage dies, lift plants and hang upside down. Shell pods when completely dry
Seedlings are not thinned, but allowed to grow thickly as sown. Fill in any gaps as seedlings appear with the extra seeds sown	2–3 in (5–7.5 cm) between plants. Distance between rows about the same as the height of plants—consult the seed pack	Protect seedlings from birds with cotton, netting, or hoop houses (remove as they grow). Hoe between rows until growth covers the ground. Water freely and mulch if soil tends to dry. Push thin twiggy supports into soil when you sow so tendrils can twine around them easily	Pick as soon as peas are swollen in the pod and use or freeze at once. Leave some peas on the plant to ripen fully for drying
No need to thin. Fill in gaps with extra plants	2–3 in x 3 ft (5–7.5 cm x 1 m)	*See* Peas	**Snow peas:** Pick when pods are 2 in (5 cm) long, before peas have formed properly. Use at once **Petit pois:** Harvest as other peas
Indoor sowings: Harden off ready for planting out in spring **Outdoor sowings:** Pull out weaker seedlings as they emerge	10 x 18 in (25 x 45 cm)	General cultivation is similar to other peas. Support emerging seedlings or leave to grow bushy over the ground	Gather when pods are about 1–2 in (2.5–5 cm) long

SUPPORTING BEANS AND PEAS AS THEY GROW

Almost all varieties of beans and peas need some kind of support as they grow, with the exception of bush or dwarf types.

1 The most common way of staking peas is to push pea sticks into the ground near the growing plants. The wispy tendrils will wrap around the sticks.

2 To support broad beans, tie string between poles that are staked at the ends of each row. The plants are then loosely tied to the strings.

3 & 4 Green beans can be grown up teepee supports consisting of four or more poles positioned on the ground in a ring or square and pulled together and tied at the top. They can also be grown along rows of poles angled together like triangles and crossed at the top. By the time of harvesting, the green beans will have reached the top of their supports.

5 & 6 Search carefully through the leaves of the plants for any concealed beans. If left, they will grow too large and become tough.

Growing Potatoes

Although potatoes take up quite a lot of space in the vegetable garden, they earn their keep. Not only are they a staple ingredient in the kitchen, they are also said to "clean" the land, especially in a new vegetable bed, leaving the soil prepared for growing other crops the following year.

The "cleaning" is actually brought about by the thorough digging that you must do before planting a potato crop, and the hilling up during the plants' growth, which helps to prevent weeds. This leaves you with well-cultivated, weed-free ground.

The different varieties of potato produce types that are round, oval, finger-shaped, or kidney-shaped. The skins may be yellow, pale brown, red, or white and the flesh may be floury (best for mashing) or waxy. Even if you have a strong preference, consult your local Cooperative Extension Office to find out the types that do well in your area.

Potatoes are divided into earlies (short season), midseason, late season, and fingerlings.

Soil and position

All potatoes do best in a fertile, well-drained, well-cultivated soil and, unlike most other vegetables, prefer acidic conditions (they prefer a pH of 5.5 or even a bit lower—*see* page 24). Ideally, double-dig the ground (*see* page 29) and amend it with organic matter, especially on light, sandy soils. Dig some bonemeal into the top layer.

Choose an open site. Heavy shading can lead to light top growth and a disappointingly small crop.

Fresh new potatoes are one of the delights of early summer. Just dig enough for a meal at a time: lightly scrubbed and steamed with their tender skins still on, they are delicious.

Planting potatoes

Potatoes are grown from seed potatoes available in late winter and early spring. Although these come from the previous year's crop, always buy them rather than saving some of your own crop, which are likely to harbor disease. For the same reason, do not use potatoes sold for eating. Instead, buy seed potatoes that are certified disease-free.

Seed potatoes should be about the size of an egg. For earlies, put

Good potato varieties

EARLIES

- **Yukon Gold** Light flesh; good all-round culinary potato
- **Sieglinde** Thin-skinned and scab resistant; good for storing

MIDSEASON

- **Chieftan** Red variety; high yielding

LATE SEASON

- **Russian Blue** Heritage potato with dark blue or purple skins
- **French Fingerling** Pink skinned, with white flesh; high yielding

CULTIVATION OF POTATOES

1 Plant individual potatoes in holes and gently cover with soil. Shoots should be facing upwards.

2 Alternatively make drills with a draw hoe and plant rows of potatoes with shoots uppermost.

3 When leaves first appear, cover bases of the potatoes with more soil and water the growing plants well.

4 When the stems are large enough, begin to hill up the potatoes at 2–3 week intervals.

5 Harvest earlies by hand, leaving the smaller crops in the soil to grow more. Lift tubers carefully so that you do not damage them with the fork.

6 Harvest midseason types when tops have died down. Leave potatoes on the ground to dry out before storing in a cool, but frost-free, dark place.

the potatoes in shallow boxes, with the eye pointing upwards. Leave in a light, warm place for a few weeks until the shoots appear—a process known as "presprouting" or "greening." Two or three will probably grow close together. Leave two shoots and rub off any others; or, if possible, cut the potatoes in half to give two pieces with two shoots, but only do this just before planting. Potatoes that are put in too warm or dark a place may develop long, spindly shoots. These will not be as healthy or easy to plant without damaging the shoots.

Make drills in the soil with a draw hoe (using the entire width of the hoe) or plant the potatoes individually with a trowel (*see* below for planting depths and times). Put the emerging shoot uppermost and pull the soil loosely over the potato; do not pack it tightly because you might damage the shoots.

CULTIVATION TIPS

Potatoes are unlike any other plant in the way in which they are planted and tended, but their needs are easily met and potatoes are usually an easy and trouble-free crop.

Immediately after planting, hoe between the lines and hill up the soil up over the tubers very slightly. This mound of earth will protect them

POTATO PLANTING TIMES, DEPTHS, AND DISTANCES

	PLANTING DEPTH	DISTANCE BETWEEN TUBERS	DISTANCE BETWEEN ROWS	WHEN TO PLANT
Earlies	3 in (7.5 cm)	10 in (25 cm)	2 ft (60 cm)	Early spring
Midseason	4 in (10 cm)	12 in (30 cm)	2 ft 4 in (70 cm)	Midspring
Late	5 in (12.5 cm)	15 in (40 cm)	2 ft 6 in (75 cm)	Late spring

Presprout seed potatoes to encourage shoots to grow. Plant them, shoots on top, and cover gently to avoid damaging the growth.

from any late frost. When the first leaves appear, either cover the plants with straw or pile some more soil lightly over them, allowing just the tips of the shoots to peep through the surface.

When the stems are 8–9 in (20–25 cm) high, start gently hilling up the soil on either side of the rows to surround the stems. This encourages the plant to develop more roots and, therefore, more potatoes. It also keeps the developing potatoes well covered in soil so the light does not reach them and turn them green. About 6 in (15 cm) of leafy growth should show above the ridge at all times. Keep the top of the ridge broad rather than narrow, and tapering. A pointed ridge makes it more likely that the potatoes will be exposed. Repeat the process every few weeks so that deep trenches develop between the rows. This process also helps to control weeds and improves drainage in the soil.

An all-purpose fertilizer can be added to the soil early in the season to help to boost your crop. In dry spring weather, make sure the potato plants are kept well-watered.

Harvesting

Start harvesting the first early potatoes in early summer. The very first harvests can be done by hand, feeling and delving into the soil and taking the largest potatoes (even these will be very small) and leaving the others to carry on growing. Make sure those left behind are well covered over with soil.

Later earlies can be dug up, using a broad-tined fork. Push this into the ground well clear of the plant, so you don't spear the potatoes. Turn the plant into the trench, then collect all the potatoes by hand.

When the earlies have run out, towards the middle of summer, start harvesting the midseason crop. They should yield good-sized potatoes until late summer.

The late potatoes should be lifted as late as possible in fall on a fine, dry day. Leave them on the surface of the ground for several hours to dry off, particularly if they are wanted for storing. Then rub off any large clumps of dirt and store the potatoes.

Clearing the plot after harvest

Put the old plants on the compost pile when you dig up your potatoes, unless they were diseased, in which case they should be thrown in the garbage. Avoid using compost made with potato plants on ground in which potatoes or tomatoes will be grown the following year, because it might still harbor harmful bacteria or viruses common to these vegetables.

At the end of each season, make sure you have removed all the potatoes from the ground. It can then be used immediately for another vegetable crop.

PESTS AND DISEASES

Potatoes are subject to a number of pests and diseases, but many can be avoided by using certified, disease-free seed potatoes. A strict crop rotation will also help prevent disease hitting your potato crop.

The most common problems to affect potatoes are flea beetles (page 144), Colorado potato beetle (page 150), potato blight (page 151), potato cyst nematodes (page 151), potato scab (page 151), wart disease (page 151), and wireworms (page 151).

Alternative growing methods

One alternative to the traditional method of growing potatoes is to place them on the ground and cover with black plastic sheeting. Anchor the edges with stones or wooden stakes.

Make slits in the plastic directly above each growing plant. Watch for slugs, which thrive in these conditions. Roll back the sheeting for harvesting.

Potatoes also do well in containers. You can use ordinary large, deep pots, planting the seed potatoes in around 8 in (20 cm) putting soil at the bottom and adding more soil on top to bury the plants again as they grow. Special potato containers allow you to add another strip of plastic at the top of the pot each time you need to top up the growing plants with dirt. Look out, too, for strong plastic bags or corrugated plastic towers and many more options for raising a crop of potatoes in a container.

Leafy Salad Greens

Salad greens are a delight of the summer and early fall months, but it is also possible to grow many of them in winter and spring too, and they can often be cooked in stir-fries as well as eaten fresh in a conventional salad. Many vegetables could be classified as salad crops, but those included here are endive, mache, land cress, cucumber, lettuce, and radishes and some of the more exotic salad greens, such as arugula, mizuna, mustards, and Asian greens like bok choy and mustards.

Endive is a useful winter salad vegetable that may be eaten raw or cooked. It is traditionally grown in two stages, first in the garden, and then in large pots. This second stage is known as forcing and it is at this time that the light-green, spear-shaped heads or "chicons" are produced.

Mache and endive also provide fresh leaves for winter salads, when other salad crops are not in such abundance, although both can be produced through the summer too. Mache has small, dark-green leaves and is a small plant; endive is much larger and may either have deeply crinkled and curly leaves or ones that are waxy and indented. Endive is blanched during growth to reduce its otherwise rather bitter taste.

Mustard and cress

Mustard and cress are usually grown indoors. Do this either by sprinkling the seed on saucers containing dampened blotting paper or sheets of paper towel, or in shallow trays of potting soil. Keep these in a dark place until the seed has germinated, then move them to a light windowsill. Cut and use when the stems are about 1½ in (4 cm) high, which will be take about two to three weeks.

GROWING TIMES AND HARVESTING MONTHS

It is possible to produce some salad greens in your garden throughout the year, although summer and fall are the principal seasons in most areas. Lettuce can be grown all year round in mild-winter climates, but they do not like hot summers.

Choose varieties to suit your taste and those you find easiest to produce. The different crops all include varieties suitable for sowing variously throughout the year.

The chart below shows the quickest time in which it takes these crops to reach maturity—remember this will vary according to the month of sowing and the weather conditions—and the months when they can be harvested.

VEGETABLE	SHORTEST GROWING TIME	HARVESTING MONTHS
Endive (curled varieties)	*To lifting*: 5 months *Forcing to harvesting:* 1 month	Early winter–late winter
Mache	4 months	Midwinter
Land cress	2 months	Late winter–winter
Cucumber	4 months	Late summer–early fall
Endive	*To blanching:* 3 months *Blanching to harvesting:* 2–4 weeks	Fall–early winter
Lettuce	2½ months	Midwinter
Other greens	4–6 weeks	Spring–early winter and All year round in mild climates
Radish *Ordinary* *Winter*	 1 month 4 months	 Spring–early winter Fall–midwinter

Land cress is much easier to grow than watercress, which should be grown in a running stream. Land cress has smaller, rather more symmetrically arranged leaves, but its flavor is very similar.

Lettuce and leaves

The two main types of lettuce—head and romaine—have countless subdivisions. As well as normal-sized lettuces in all different shades of green, there are dwarf varieties, those with very crisp leaves that curl over each other and others with soft leaves that form loose hearts. Choose according to the type you like and the time of year in which you want to grow them. There is always a wide selection of lettuce seeds in garden centers in early spring.

Many salad leaves are grown as cut-and-come-again crops. They grow as individual leaves, which are picked as required. More grow in their place, so that you can continue to harvest over a long period. You could grow a dedicated crop of arugula, for example, or choose a mesclun seed mix that has a variety of greens, often including arugula, endive, mizuna, mustards, and more colorful varieties, such as Bull's blood beet and red salad bowl. Spicy mizuna mix is a widely available and popular choice, or you can experiment with creating your own mixtures.

Bok choy is harvested like a hearted lettuce, but is included with cut-and-come-again leaves here, as it requires the same care.

Other salad ingredients

Radishes are actually cole crops, but are included here as they are grown and eaten as a salad crop. Ordinary radishes grown throughout the summer may be round, bomb-shaped, or conical, and brilliant red, red and white, or even pure white. Winter radishes are parsnip-shaped, black-skinned and can grow up to 6 in (15 cm) long.

In coldest areas, cucumbers are often grown as a greenhouse crop, as they need warm conditions to grow. Different types include those for slicing, pickling, and "burpless."

Tomatoes are such a varied salad crop that they are covered separately in greater detail on pages 87–90.

CULTIVATION TIPS

The comprehensive cultural details for the leafy greens in this section are listed overleaf. Nearly all these crops do best in a fertile soil that is rich in organic material.

This need for fertile soil is particularly true of cucumbers, which can be successfully grown on a mound of well-rotted compost or manure covered with 3–4 in (7.5–10 cm) of good topsoil. The roots of cucumber plants need to be kept very moist, but the crop will fail if they are waterlogged. Cucumber plants can be left to trail over the ground (but remember they take up quite a lot of space), or trained to grow up a vertical trellis.

Greens as intercrops

As most salads are quick to mature, they can be sown as intercrops—that is, between rows of slower-maturing vegetables. Alternatively, lettuces can easily be grown in odd spots in a flower garden or even in containers or hanging baskets. Sow all types of greens very thinly and in rapid succession. As leafy greens cannot be stored, it makes more sense to have a small, steady flow, rather than to be swamped with a glut of say, radishes, and lettuce, all at once.

Most greens need to be kept well watered while growing. Most must be grown rapidly both for the best taste and to stop them bolting. Once harvested, pull up the roots quickly and compost them. The land can be used immediately for another crop.

Add color and interest to your salads when you grow your own crops. You need not be restricted to the bland green varieties available in the supermarket when you can raise your own radishes (above left) *and colorful leaves like radicchio, Bull's blood beet, and red salad bowl lettuce* (above).

Winter salads

Endive, radicchio, mache, mustards, and many Asian greens are cool-season vegetables, but they can be sown in early spring to crop in summer in areas that are not too hot. Seed catalogs increasingly offer a range of other winter-hardy greens such as escarole and Italian dandelion.

Winter crops may need to be protected from harsh weather with row covers or hoop houses and even though weeds are not as abundant at this time of year, it is still good practice to hoe regularly to keep them at bay.

CULTIVATION OF SALAD GREENS

VEGETABLE	SOIL/POSITION	SOWING	DRILL DEPTH
Asian greens	**Soil:** Any fertile, well-drained soil; pots of all-purpose potting soil **Position:** Open and sunny	Sow in spring for early summer crops and again in fall for late fall and winter crops	¼ in (0.5 cm)
Endive (curled varieties)	**Soil:** Deep and well-cultivated. Appreciates some well-rotted manure or compost dug in deeply before sowing **Position:** Any, but does not like to be heavily overshadowed	Sow thinly in cropping position from spring to early summer	¼ in (0.5 cm)
Cress, land	**Soil:** Damp, full of well-rotted manure **Position:** Moist and shady	Water the drills well and sow seed thinly in spring; then again in fall	¼ in (0.5 cm)
Cucumber	**Soil:** Good soil that has been well enriched with manure or compost shortly before planting **Position:** Sunny and sheltered	Sow indoors in individual peat pots in late spring. Keep in warm place. Sow outdoors when all danger of frost has passed and the soil has warmed. Sow three seeds in a 4–6 in (10–15 cm) group, every 2 ft (60 cm)	½–1 in (1–2.5 cm)
Endive	**Soil:** Light, well-drained and fertile soil **Position:** Any, but prefers a well-sheltered spot	Sow outdoors in spring; cover with floating row covers to warm soil	½ in (1 cm)
Lettuce	**Soil:** Fertile, well-drained light soil. Crisp varieties like soil with lots of organic matter incorporated to help make them water-retentive **Position:** Open and sunny. Will not generally grow well if trees are shading the site	**Harvesting times:** Depending on your region, lettuce varieties can be sown year round. In many regions early spring and late summer are best for direct sowing. In warmer areas, lettuce can be sown throughout the winter months	All drills ½ in (1 cm)
Mache (Lamb's lettuce)	**Soil:** Any well-cultivated soil **Position:** Open and sunny	Sow thinly in early spring and again in late summer	½ in (1 cm)
Radish	**Soil:** Any fertile, well-cultivated soil, preferably containing well-rotted organic matter. This is particularly necessary if soil is light and sandy **Position:** For spring sowing: open and sunny, but sheltered. Overshadowing may result in radishes that are woody and bitter. For summer sowing: cool, shady site	**Summer radish:** Sow outdoors from late winter to late summer; sow very thinly every two weeks. If you want radishes even earlier (for late March), sow under cloches or row covers in January/February. Put cloches or covers in place a few weeks before sowing to warm ground **Winter radish:** Sow outdoors in late summer. If ground is dry, water base of drill before sowing	½ in (1 cm) ¼ in (0.5 cm) ¾ in (2 cm)

THINNING/TRANSPLANTING	PLANTING DISTANCES	CULTIVATION	HARVESTING
If sown thinly, no thinning should be necessary. If you need to thin out the crop, pull seedlings when they are large enough to eat	8–10 in (20–30 cm) between rows	Water freely and hoe between rows. Cover early crops and late sowings with cloches or row covers in cold regions if needed. Some will bolt if left to grow for too long or if allowed to dry out	Pick leaves as required
Thin progressively to recommended spacing	8 x 15 in (20 x 40 cm)	Hoe between rows, and water if plants show signs of drying. In late fall, when the leaves have turned yellow and died down, dig up the long roots. Cut off leaves at about 1 in (2.5 cm) and rub off all the lower end of the root to make it about 8 in (20 cm) long. Force at once (*see* Harvesting) or store covered in dry soil in a dark place until wanted	**To force:** Plant roots in containers filled with potting soil, so there is 1–2 in (2.5–5 cm) between them and the tops are visible above the soil. Water and cover container with another pot the same size, inverted. Put in a warm place and leave for about four weeks. The heads are ready when they are 5–6 in (12.5–15 cm) high. Cut as required and then compost the roots
Thin progressively to recommended spacing	8 x 12 in (20 x 30 cm)	Water freely. Mulch the ground if there is any likelihood of it drying out	Pick outer leaves first. As these become tough, use the center leaves
Harden off those grown indoors in late spring. Plant out in early summer, spacing as recommended. If sown outside, pull out the weaker seedlings as they emerge, leaving one strong plant at each spot	2 x 3 ft (60 x 90 cm)	Water freely. It helps to encourage sideshoots if you pinch out the growing tips when four or five leaves have formed. Pollinate female flowers by pushing pollen-bearing male flowers into their centers or using a small brush. As young cucumbers form, pinch out shoots that have no cucumbers beyond the seventh leaf	Cut the cucumbers while they are still young. More will form
Thin progressively to recommended spacing if sown outdoors	12 x 15 in (30 x 40 cm)	Hoe to eliminate weeds and water freely. Give a liquid fertilizer to encourage growth if necessary. Blanch center leaves after about three months (*see* page 86). Alternatively, cover the row of endives with black sheeting or white-painted cloches	If harvesting in fall, plants will take 10 days to blanch. In winter, leave them for twice this time. Lift all plants before any danger of frost. They will keep for a few weeks in boxes of soil in a dark shed or garage, but are best used immediately
Prick out seedlings from early sowings. Thin all crops as large enough to handle, working progressively to recommended distance, transplanting second thinnings and eating third thinnings as appropriate	From 6–12 in (15–30 cm) between plants: check seed pack. Distance between rows is usually 12 in (30 cm)	Protect all seedlings from birds with threads of black cotton, netting, or row covers. Summer crops need constant watering: they will bolt if allowed to dry out. Winter lettuces need less watering, but must be sown in well-cultivated soil. To encourage romaine lettuce to form its center heart, draw up the outer leaves and tie loosely around the plant	Head lettuces are ready when the heart is well formed. Feel it gently, but do not pinch it. Crisper varieties are ready when the center leaves have curled over each other and this part feels hard and compact. Romaine types are ready when the center leaves are well developed. Cut all lettuces as required, but do not allow them to bolt
Thin to recommended spacing	4 x 6 in (10 x 15 cm)	Water freely and hoe between rows to eliminate weeds. Protect from frost by packing straw round plants or protect with cloches or row covers	Begin to harvest when at least six leaves have formed. Pick older leaves first or pull up the whole plant.
Summer radish: Little thinning will be necessary, assuming seed was thinly sown **Winter radish:** Thin to recommended spacing as seedlings emerge	**Summer radish:** 1 x 9 in (2.5 x 25 cm) **Winter radish:** 8 x 12 in (20 x 30 cm)	**Summer radish:** Water well—crops depend on rapid growth. In dry conditions, radishes will be woody and very hot-tasting. Summer-sown crops are prone to very rapid bolting. Cover with shade cloth in very hot weather **Winter radish:** Hoe between rows to eliminate weeds	**Summer radish:** Pull largest ones first (you can check by pushing aside leaves and stems), beginning when they are ½ in (1 cm) in diameter. Pull up in succession quickly **Winter radish:** Start harvesting in November. Can be left in the ground through winter, but are better stored like other root crops (*see* page 207)

Pick salad greens little and often. They will not keep well and are far better eaten when they are absolutely fresh. Many salad leaves will grow again to replace what you have cut.

PESTS AND DISEASES

Their fast growth and prompt harvesting means that salads are generally less affected by pests and disease than many other crops.

Slugs (page 147) will chew their way through leaves. Look out, too, for cutworms (page 147) and downy mildew (page 146). Radishes, which are cabbage family plants, will be subject to the cole crop diseases (*see* page 65); flea beetle is the worst. Do not grow radishes in ground infected with clubroot. The most common problems to look for are gray mold (page 146), mosaic virus (page 146), leaf miners (page 148), and thrips (page 150).

FORCING ENDIVE

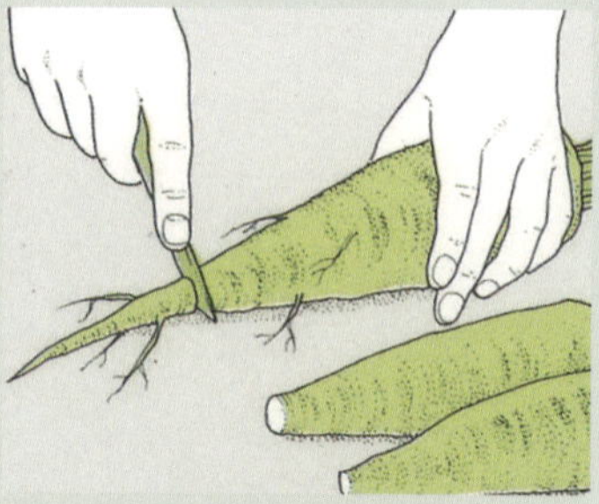

1 Endive requires two growing periods. When the long roots are dug up they are trimmed of any shoots, topped and tailed.

2 Replant in potting soil, keeping the crowns just above the soil. Water and cover with black plastic or another pot to blanch the new chicons.

BLANCHING ENDIVE

1 Loosely tie raffia around the stems of fully-grown plants. This keeps the lower tresses off the ground and reduces the risk of them rotting.

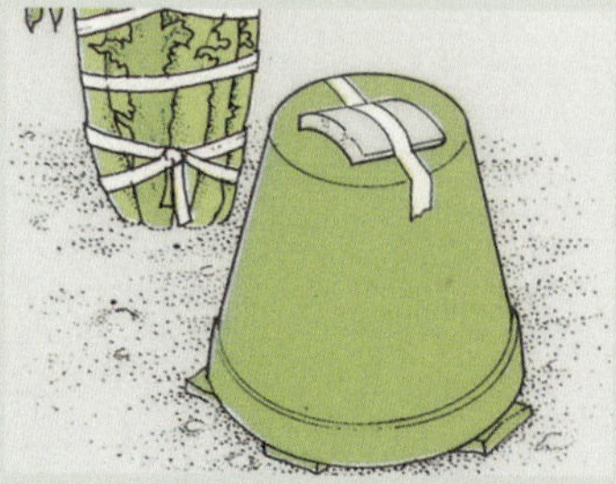

2 Cover the drainage holes of flower pots to exclude light and place over plants to blanch the inner stems. Allow ventilation around the base of the pot.

ENCOURAGING CUCUMBERS

1 Pinch out any growing shoots when enough cucumbers have formed on the plant, to prevent any further crops growth.

2 If necessary, hand-pollinate the female cucumber flowers, either with a brush or by pushing the male flowers into the centers of the female flowers.

Growing Tomatoes

The many varieties of tomatoes available today have made it possible to grow tomatoes outdoors even in northern gardens. However, a good-sized crop can often be produced on a deck or patio, on a windowsill, or in a greenhouse or sunroom.

The varieties of tomatoes are endless, from round ones of all sizes—those that are hardly bigger than a marble to those that are almost too big to hold in the hand—to plum- or pear-shaped ones that may be yellow, striped green and red, or yellow and red.

The most typical way of growing tomatoes outdoors is on a tall, single-stemmed vine, but there are also bush and dwarf varieties, the latter growing to a height of no more that 6 in (15 cm) but sprawling extensively over the ground. Many catalogs and growers refer to tall types as "indeterminate" and the more compact, bushy tomatoes as "determinate." Cherry and pear varieties can also be successfully grown in hanging baskets, where they will drape over the edges and there are many specifically bred for this situation.

Soil and position

For tomatoes, the ground should be well-cultivated and well-drained soil, into which well-rotted organic matter (leafmold is ideal) has been incorporated the previous fall. If the ground is freshly amended just before planting tomatoes, the plants will be very leafy but bear few fruits. Tomatoes will grow well in potting soil and grow bags, although they will need regular liquid fertilizer in the later stages of growth.

Sunshine is the main requisite for a good crop of juicy tomatoes, so choose the sunniest spot you can.

Tomatoes are America's favorite garden vegetable for a good reason. Homegrown fruits have much more flavor than store-bought.

South-facing walls or fences make the best sites, because they also act as windbreaks.

SOWING AND PLANTING

Tomatoes can be grown successfully from seed and there are many dozens of varieties to choose from. Make sure you buy a variety the right size for your garden.

Sow tomatoes in cell packs or pots filled with growing mix in early spring. Sow seeds 1 in (2.5 cm) apart or two seeds to each 3 in (7.5 cm) pot. Keep the soil moist and cover with newspaper until the seed has germinated. Put the containers in a warm, darkened place and, as soon as the seedlings emerge, remove the newspaper and move the containers to a warm, south-facing windowsill in full sunlight. When the first leaves appear, transplant strong seedlings to individual pots filled with potting soil, or just remove the weaker seedling of the two if they were sown in individual pots.

Good tomato varieties

- **Brandywine** Very popular heavy cropper with large fruits
- **Moneymaker** Popular old English variety suitable for greenhouses
- **Green zebra** Medium-sized fruit lime green streaked yellow; tart
- **Sungold** Sweet, cherry-sized fruits that are golden in color
- **Sweet Million** Tall, with long trusses or small, sweet red fruits
- **Yellow Pear** Tall, heirloom variety; late-ripening with small fruits
- **La Roma** Heirloom bush type with early yields of Roma tomatoes

CULTIVATION OF TOMATOES

1 Sow the seed in trays filled with growing mix in midspring. Dampen the mix, cover the trays with newspaper, and leave them in a warm, dark place.

2 When the seedlings begin to grow, prick out the strongest and plant in individual flower pots filled with potting soil. Water the seedlings lightly, but regularly.

3 Harden off the seedlings either by placing them in cold frames in midspring, or under hoop houses in late spring. If necessary, repot any seedlings that are outgrowing their pots.

4 Plant out the seedlings when they are about 8 in (20 cm) high. Tip them out of their pots, holding them carefully by their rootballs, and plant them in the soil.

5 Space plants 18 in (45 cm) apart. Tall varieties will need to be supported with cages or canes, loosely attached with soft garden string.

6 Water the plants regularly to prevent them wilting. Take care with plants in grow bags, as too much water can rot their roots.

7 Give a regular liquid fertilizer from midsummer, once they begin to bear fruit. Irregular feeding and watering may make the fruits grow unevenly and split.

8 Remove any yellow leaves and tie back others that are shading tomatoes from the sun. Place straw on the ground to protect lower trusses from dust and dirt.

Healthy tomato seedlings will have a strong stem and deep green, robust leaves; discard any that are weak and straggly. It takes about seven weeks from sowing the seed to the plants being ready to plant out into the garden.

From midspring, begin to harden off the plants by putting them in cold frames. If you don't have one, put them under hoop houses in late spring, or keep indoors until early summer. If necessary repot the seedlings into bigger pots as they grow. Keep the potting soil damp, but not waterlogged.

Planting out established plants
Many people prefer to buy plants from a nursery rather than raise their own from seed. Buy them when they are about 8 in (20 cm) tall, making sure you buy sturdy-looking plants. They should be dark green and short-jointed; avoid those that look weak or have large gaps between the fernlike leaves.

Plant tomatoes carefully; tip them out of the pot and hold them by the rootballs, not the stems. Plant with the rootball and soil intact, so the top of the rootball is at least 1 in (2 cm) below the surface of the soil. Plants should be about 1 ft 6 in (45 cm) apart. If growing more than one row, place rows 2 ft 6 in (75 cm) apart.

Growing tomatoes in pots and containers

Planting in containers means that you can position your tomatoes in the sunniest, most sheltered spot.

Containers
Any good-sized container will house a tomato plant, although you will need to be vigilant because pots dry out faster than garden soil. Make sure that pots are sturdy and have a broad base, as the tall plants and heavy fruit make tapering pots unstable.

Hanging baskets
Cherry and pear tomatoes are the best choice for hanging baskets, as the fruit will not get so heavy that it risks breaking the plant's stems. Add slow-release fertilizer and water-retaining gels to reduce the need for feeding and watering. "Tumbler" tomatoes are best for baskets.

Grow bags
Some are formulated specifically for tomatoes and you may not need to fertilize often. Plants will still need staking; look for wire frames that slide beneath the bag and curve over the top, and poke canes through the supports to make it more secure. Grow bags are most often used in cooler climates so that tomatoes can be grown inside a greenhouse.

Other gadgets
A wide variety of devices can be found in catalogs to help you grow tomatoes. Tomato cages range from colorful metal surrounds to hinged steel cages that can be staked to the ground. Adjustable rings or spirals must be secured to a stake and can be adjusted as the plants grow. Upside-down cylindrical grow bags are meant to be hung on a patio or deck. Another popular device is a "Wall O' Water," which is placed around young transplants and filled with warm water to warm the plants.

CULTIVATION TIPS

Tomatoes need food and water throughout their growing period. Irregular watering can lead to split fruits and a poor harvest.

Water the seedlings immediately after planting. A good way to keep tomato plants evenly moist is to sink a 4 in (10 cm) flowerpot into the soil near the root and fill this to the top with water once a week. This allows water to get straight to the roots and to seep into the soil slowly and gradually.

Push cages or strong canes about 5 ft (1.5 m) tall into the ground close to the plant and as the plant grows, tie the stem loosely to the cane (giving the stem room to expand without damage from the ties).

Pinching out
As growth progresses, pinch out the sideshoots that grow in the leaf axils—that is, the little shoots appearing between the leaves and the stem (*see* page 90). This helps to ensure that strong flower trusses (on which the fruits will form) develop, instead of a number of long, useless sideshoots. The flower trusses grow from the main stem, not the leaf axils and these unnecessary shoots will take up valuable water and nutrients.

When the plant has developed about four trusses (usually towards the end of July), prevent further upward growth by pinching out

Every two weeks, pinch out any shoots growing in the leaf axils to prevent them from taking valuable nutrients from the developing fruit.

the central growing tip. The plant's energies will then be directed into developing the trusses. Pinching out is not necessary on most cherry or bush tomato varieties.

Tomatoes will benefit from regular liquid fertilizer from now on; you will need to begin earlier if growing in containers or grow bags, as the fertilizer in the potting soil will soon be exhausted by the growing plants. Spray the plants with liquid tomato fertilizer, as well as watering through the ground to the roots.

Good housekeeping, healthy plants

Remove any leaves that turn yellow as soon as you spot them and, as the fruits begin to form, tie back any lower leaves that are shading them from the sun. Plants with a lot of leaf growth can become prone to disease, as air cannot circulate within the plant and moisture builds up, encouraging disease. Tomatoes on lower trusses can be kept clean by putting a layer of straw beneath them to protect them if their stems bend down to the ground.

Harvesting

Begin picking the tomatoes as soon as they are ripe, supporting the fruits in the palm of your hand and pressing the stem just above the fruit. It will snap, leaving the calyx on the end of the tomato.

In fall, any fruit still remaining should be picked, even if it is green—it will not survive early frosts and will be lost entirely if left too long. Pull up the plant and hang either the whole thing, or just the tomato-bearing trusses in a warm, preferably sunny, spot indoors, or in a greenhouse, to allow the tomatoes to ripen. Alternatively, if space permits, leave the plants in the ground, but untie them from their supports, lie them down gently on a bed of straw, and cover with row covers until the tomatoes ripen.

Green tomatoes can also be picked off the plant and put carefully into trays, placed in a cool, darkened place. They will ripen gradually over the next few months, and will be encouraged to do so if you include one red tomato with every batch. Alternatively, use the green tomatoes to make chutney, salsa, or relish (*see* pages 214–17).

PESTS AND DISEASES

The tomato is closely related to the potato and therefore is likely to suffer from many of the same problems. Plants may be affected by the potato cyst nematode (*see* page 151) and may get blight if you handle tomato plants after tending potatoes. Always wash your hands between the two tasks.

Apart from those diseases mentioned below, young plants put out too soon may grow slowly and look sickly; protect them with row covers or cloches. Tomatoes grown outdoors, however, are usually far less susceptible to problems than greenhouse equivalents.

The most common problems are blossom end rot (page 150), virus (page 150), gray mold (page 150), and splitting (page 150).

HARVESTING TOMATOES

1 Pick tomatoes by supporting the fruit in the palm of your hand and gently snapping off the stem above the calyx.

2 If picking whole trusses of green tomatoes at the end of the season, leave them outside on a bed of straw under row covers or a hoop house to ripen.

3 Alternatively, the trusses can be cut and hung up in a warm, sunny spot indoors, or ideally in a greenhouse, until they are ripe.

Other Vegetable Crops

Included on the following pages are some of the many other vegetables you may choose to cultivate, including artichokes, Jerusalem artichokes, asparagus, celery and celeriac, leeks, onions and garlic, squash, zucchini, pumpkins, okra, spinach, and corn.

There are two main types of celery—ones that must be blanched during growth and others that are self-blanching. Self-blanching celery is easier to grow and demands less work, but it will not survive the heavy frosts that the other types can withstand and is generally said to be less tasty. Celeriac, known sometimes as the turnip-rooted celery because of the edible swollen bulb that grows at the bottom of the plant, is easier to grow than celery and can be eaten raw or cooked.

Artichoke varieties

Artichokes are perennial plants, and their silvery green leaves make them an attractive addition to the garden. Each plant yields about six flower heads (the edible part) a year and will crop for about six years, although it is best to renew plants after three or four years, as the heads start to become smaller and tougher. You probably don't need more than a few plants, but growing your own will give you more opportunities to eat this delicious vegetable, which is often considered to be something of a delicacy.

Jerusalem artichokes are grown for their tubers, which have a distinctly earthy taste. The plants resemble sunflowers (to which they

Artichokes are a delicious treat for the table, but worth growing for their striking beauty, too. These stately, silvery plants bear fantastic purple flowerheads in summer.

GROWING TIMES AND HARVESTING TIMES

The chart below gives the shortest time it takes for these other crops to reach maturity, and the months in which they are generally harvested. As with other vegetables, cropping seasons can be extended by regular sowings at intervals of a couple of weeks or a month, or by sowing different varieties that take varying times to reach maturity.

VEGETABLE	SHORTEST GROWING TIME	HARVESTING TIMES
Celeriac	6½ months	Late fall–early winter
Celery:		
Self-blanching	5½ months	Late summer–midfall
Blanching	6½ months	Early fall–late winter
Corn	3 months	Late summer–fall
Leek	8 months	Midfall–spring
Okra	3 months	Late summer–late fall
Onion	6–10 months	Fall–winter
Pumpkin	3 months	Late summer–late fall
Scallions	3 months	Late summer–early winter
Shallot	4 months	Late summer
Spinach:		
Summer	2 months	Spring–fall
Winter	2½ months	Early winter–spring
New Zealand	1½ months	Midsummer–fall
Squash	3 months	Late summer–late fall
Zucchini	2½ months	Late summer–early fall

are related), and they grow tall enough to form an effective screen to hide a shed or compost pile. They will grow in almost any soil.

Squash and pumpkins

In addition to the familiar green and yellow zucchini and old favorites like orange pumpkins, there are a great variety of vegetables in this group, from delicatas to English marrows, from decorative gourds to acorn, spaghetti, crookneck, and scallop squash. There is a huge range of colors and shapes, bush and trailing types, and most squash and pumpkins even have flowers that are themselves edible.

Plants are typically divided into two categories: summer squash and zucchini are grown during the hot summer months. Winter squash and pumpkins need a long season to ripen, so you must time your sowing or planting to give them the maximum amount of time outdoors.

The cultivation for all kinds of both summer and winter squash is the same. Remember that these can be sprawling plants that require plenty of room to spread. If your garden is on the smaller side, look for trailing varieties, which can be trained to grow up vertical supports if space is at a premium.

Ornamental gourds are grown in the same way, although most types are purely decorative and not edible. For eating, pumpkins are usually not as tasty as summer or winter squash, but can be used in many dishes, including traditional pumpkin pie, soups, stews, and more.

The onion family

Leeks, onions, scallions, and garlic are all members of the same family, but leeks are the easiest to grow and the hardiest, as they will withstand heavy frosts. Growing them also helps to break down compacted soil and improves its texture for other crops the following year.

Onions can be produced the whole year round in some regions, but they respond to the number of daylight hours, which varies greatly from north to south. In general, short-day varieties are best for southern gardens and long-day onions are best for northern gardens. In mild-winter areas, short-day types can be grown through the winter.

They are grown from onion sets, small bulbs planted into the ground—a more successful and higher-yield method than trying to grow from seed. Each set swells into one large bulb for harvesting. When the leaves turn yellow, bend them over the bulbs and leave the onions to finish ripening. Once they are harvested, they can be dried by hanging them in bunches or braids.

Okra, spinach, and corn

Okra (originally from Africa) is an old Southern favorite that can be used to flavor soups, stews, and curries, and it can also be served as a vegetable side dish. It is also known as ladies' fingers or gumbo. It grows best in hot, humid conditions.

Harvest spinach leaves when they are young for their best flavor. Larger leaves develop woody, stringy stems and tough foliage.

Okra needs hot temperatures to thrive, but can be grown in northern gardens if it is started indoors to get a jump on the season.

Spinach is one of the fastest growing of all leaf vegetables, making it a good intercrop vegetable when you have an area of empty ground. The two main types are smooth-leafed and savoy-leafed spinach, but New Zealand spinach is usually grouped with them, although it is not a true member of the spinach family. It has smaller leaves that are less shiny, and it thrives in full sun and warm conditions. New Zealand spinach will flourish in drier conditions, so is a useful alternative in sandy, free-draining soil. Perpetual spinach is included with root crops (*see* pages 66–71).

Corn is traditionally a hot-weather vegetable, but there are plenty of varieties that can be grown in more temperate climates.

CULTIVATION OF OTHER VEGETABLE CROPS

VEGETABLE	SOIL/POSITION	SOWING	DRILL DEPTH
Artichoke	**Soil:** Rich, well-drained soil that was amended the previous spring. Apply lime about two months before planting if soil is acidic **Position:** Open, sunny, but sheltered	If raising from seed, sow outdoors in spring. If raising from divisions (also called shoots or suckers), plant them outdoors in late spring	½ in (1 cm)
Celeriac	**Soil:** Richly cultivated and well-drained. Will grow in worse soils than celery, but bigger vegetables are produced in rich soil **Position:** Sunny	Sow indoors in midspring and keep warm until germination	¼ in (0.5 cm)
Celery	**Soil:** Deeply dug, rich, well-drained soil with lots of well-rotted organic matter. Self-blanching types will grow in poorer soils **Position:** Open and sunny	**Both types:** Sow seeds indoors at two week intervals in spring to stagger the crop. Keep in a warm place to germinate	¼ in (0.5 cm)
Corn	**Soil:** Well-cultivated, rich soil. Improve light soils by incorporating lots of well-rotted organic matter **Position:** Warm, sunny, and sheltered, but not shaded	Sow outdoors in late spring. Plant two or three seeds at each planting hole (*see* Spacing). Plant at two-week intervals for a longer season	½–¾ in (1–2 cm)
Garlic	**Soil:** *see* Onions **Position:** *see* Onions	Divide garlic bulbs into individual cloves and plant, growing tip upwards, just below the soil at intervals in spring and again in fall	*See* Onions
Jerusalem artichoke	**Soil:** Will grow in almost any soil; the richer it is, the heavier the crop. Amend poor soil with compost the winter before planting **Position:** Sunny, but will tolerate partial shade. Can be grown as a screen	Sow indoors in spring or plant tubers outdoors late winter to early spring	4–6 in (10–15 cm)
Leek	**Soil:** Well-cultivated and well-drained soil, which was amended with compost or well-rotted manure the previous winter. Leeks are not as demanding as onions in their soil requirements **Position:** Any, but they dislike heavy shade	Sow indoors in late winter. Outdoors in early spring, sowing three seeds every 1 in (2.5 cm)	½ in (1 cm)

THINNING/TRANSPLANTING	PLANTING DISTANCES	CULTIVATION	HARVESTING
Thin seedlings as they appear, discarding the weaker ones. The following spring, transplant into permanent site at recommended spacing	2 ft 6 in x 3 ft 6 in (75–105 cm)	Mulch plants one month after planting out. Water well in dry spells. Remove flower buds as they appear in the first year. In subsequent years, allow plant to develop four to six flower buds, picking off any extra ones. In early winter, cut main stems to ground level, leaving small leaves at the center to protect the plant. Cover the plants with straw or mulch them	In the second and following years, pick the heads in summer when they are plump, but green, and the petals are tightly grouped. There should be no brown tips to the scales. Use pruners to cut them and begin with the central, largest, head. After harvesting, cut the stems back by half
Treat as celery seedlings, pricking out and hardening off for planting out in late May/June. Plant directly into the cropping site with roots well buried so plants are firm in the soil	12 x 8 in (30 x 20 cm)	Hoe to control weeds and keep well watered. Give a liquid fertilizer every two weeks. From late summer, remove any side growths and yellowing leaves to ensure the roots develop to their full potential	Begin lifting when roots are large enough (usually late fall), as required. Harvest the entire crop by winter and store any that are not required for immediate use
For blanching: In spring, dig trenches 15 in (40 cm) wide and 1 ft (30 cm) deep. Make them 1 ft (30 cm) wide for a double row of celery and plant in pairs. Dig compost into the bottom to leave trench 6 in (15 cm) deep. Plant seedlings in the trench at recommended spacings and time **Self-blanching:** Plant in a block rather than rows, so inner plants are blanched naturally	**For blanching:** 6 in (15 cm) between plants **Self-blanching:** 9 x 9 in (25 x 25 cm)	Water all plants generously after planting and during dry spells **For blanching:** When plants are 1 ft (30 cm) high, tie paper around their stems, water and hill up dirt around stems to exclude light. Make sure the leaves remain exposed. Continue hilling up regularly for six to eight weeks **Self-blanching:** Remove side growths as they appear; when plants are 1 ft (30 cm) high, pack straw around the outer stems	**For blanching:** Harvest from late summer; flavor is improved after the first frost. Dig up plants from one end of the trench, shake off soil and cut off roots. Make sure those left for later harvest are well hilled up **Self-blanching:** Harvest from end of summer until the first frost. Pack straw around plants exposed after lifting others. Harvest as for blanching types
Pull out weaker seedlings of as they emerge. Grow corn in a block rather than in rows. This gives a better chance of good pollination	1 ft 6 in x 2 ft (45 x 60 cm)	Hoe to control weeds and hill up soil up around stem to firm plants in. Water well, and feed weekly with liquid fertilizer. Remove sideshoots at the base of the stem. When three cobs have formed, remove any others that appear	When the hairy tassels protruding from the top of the cobs turn dark brown, test the grains of the cob for ripeness. A milky liquid should ooze out when you press a grain with a fingernail. Grains should be a pale yellow color. Snap off the cobs and use immediately or freeze
No thinning necessary	6 x 12 in (15 x 30 cm)	Hoe to control weeds and water well in dry spells. When flowers appear, feed fortnightly with liquid fertilizer	Harvest when leaves turn yellow. Those planted in March will be ready from October. Those planted in October will be ready the following July/August
Prick out into trays when large enough to handle. Harden off seedlings from early sowings in midspring and plant out in late spring. Harden off and plant out later sowings in early summer	1 ft 3 in x 2 ft 6 in (38 x 75 cm)	Hoe to control weeds and pinch out growing tips to prevent flowers forming. Stake plants if necessary and cut the stems down in early winter	You can leave the crop in the ground or dig them up and store. Save some tubers from the crop to plant out the following year
Thin to about 2 in (5 cm) apart when large enough to handle **To transplant:** Trim top third of leaves with scissors. Make holes in ground and drop in leeks so tips of leaves just show. Do not firm soil, just water them well. This will firm them in sufficiently	9 x 12 in (25 x 30 cm)	Hoe to control weeds and hill up soil around stem as plants grow. This helps to blanch the stems, but try not to let the soil go in between the leaves	Dig up as you need them through fall and winter, teasing them gently out of the soil with a fork. Lift in spring when you want the ground for another crop, and put in a shallow trench in a patch of spare ground. Cover roots and stems with soil, with leaves protruding. Use as required

CULTIVATION OF OTHER VEGETABLE CROPS

VEGETABLE	SOIL/POSITION	SOWING	DRILL DEPTH
New Zealand spinach	**Soil:** Light and well-drained **Position:** Sunny	Sow indoors in late winter Sow directly into cropping site in late spring Sow two seeds together every 6 in (15 cm) along the row	1 in (2.5 cm)
Okra	**Soil:** Light, which has been well-cultivated to ensure good water retention **Position:** Warm and sheltered. A south-facing spot is ideal	Sow indoors in spring. Sow direct into cropping site in late spring, but protect with row covers or hoop houses	½ in (1 cm)
Onion No onion seed or seedlings should be sown or transplanted until the soil has really dried out after winter. If it sticks to your boots, it is still too wet. Firm the ground and rake it to a fine tilth	**Soil:** Well-cultivated, deeply dug soil which has been enriched the previous fall. Scallions are better grown in lighter, not so rich, soil or they will grow too big **Position:** Sunny and open	**Scallions:** Sow outdoors at monthly intervals from spring through fall **Onions from seed:** Sow indoors in early winter. Sow outdoors in early spring **Onions from sets:** Trim off wispy growth and push sets into the soil at the recommended distance in spring	½ in (1 cm) ¼–½ in (0.5–1 cm)
Pumpkin	**Soil:** Very rich with lots of well-rotted organic matter incorporated **Position:** Warm and sunny. Will grow in light shade, but the vegetables will be smaller	Indoors, sow two to a cell pack in spring. Outdoors, sow directly in midspring, protecting young seedlings with row covers or hoop houses	1 in (2.5 cm)
Spinach	**Soil:** Rich, moisture-retentive soil (this is particularly important for summer spinach) **Position:** Summer spinach: full sun or partial shade; Winter spinach: sheltered site	**Summer spinach:** Sow outdoors during spring and late spring **Winter spinach:** Sow outdoors in late summer and early fall	1 in (2.5 cm)
Summer squash and zucchini	**Soil:** Rich and moisture-retentive. It should have lots of well-rotted manure added to it **Position:** Sunny or partial shade. Can be grown on a compost pile; trailing varieties can also be grown up and along a fence	Indoors, sow two to a cell pack in midspring. Outdoors, sow directly into the cropping site in later spring, but protect with row covers in cold areas. Sow seeds at 6 in (15 cm) intervals, the pointed end pushed down into the soil	1 in (2.5 cm)

THINNING/TRANSPLANTING	PLANTING DISTANCES	CULTIVATION	HARVESTING
Prick out indoor seedlings when large enough to handle. Harden off in late spring and plant out in early summer. Remove the weaker seedlings of the outdoor sowing to achieve recommended spacing	1 ft 6 in x 2 ft (45 x 60 cm)	Water plants constantly through any dry spells. Pinch out the central growing tips. This encourages the growth of sideshoots that bear more leaves	Pinch off young shoots when they have two or three good-sized leaves. As with summer and winter spinach, do not strip the plant
Prick out indoor seedlings into pots as they emerge. Harden off in late spring and plant outdoors in early summer. Thin those sown in outdoors progressively until you reach the recommended spacing	1 ft 8 in x 2 ft (50 x 60 cm)	Hoe to control weeds and keep plants well watered. Give plants feeds of liquid fertilizer every two weeks when the flowers have appeared	Cut pods when 6–8 in (15–20 cm) long. Regular picking encourages more growth. Plants will be killed by the first frost, so harvest the crop before then. Freeze any surplus
Onions from seed: Prick out when large enough to handle. Harden off in midspring and plant in cropping site in spring Outdoors, thin to 2 in (5 cm) when large enough to handle. Thin a month later to 4 in (10 cm), using thinnings for salads or small onions. Thin to final spacing a month later **Scallions and onions from sets:** No need for thinning	**Scallions:** 9–12 in (25–30 cm) between rows **All others:** 6 x 12 in (15 x 30 cm). If growing a large-bulbed variety, increase distance between bulbs to 9 in (25 cm)	**Scallions:** Hoe to control weeds and protect fall sowings with row covers to extend the season **All others:** Hoe to control weeds. Keep well-watered, and if any plants produce flower heads, pull them up and use the onions straightaway. When plants are fully grown and leaves begin to turn yellow, bend the leaves over just above the bulbs or allow them to topple naturally	**Scallions:** Pull when they are about 6 in (15 cm) high (the size of a pencil) **All others:** A week or two after bending over leaves, dig up onions gently to break their roots, but leave them in the ground for another two weeks. Then dig up completely and leave in the sun to ripen. Store as described on page 208. Onions for winter storage should be lifted in fall
Pull out weaker seedlings of those in cell packs as they emerge. Harden off for planting out in late May/early June. Thin outdoor seedlings to final recommended spacings	3 x 4 ft (90 x 120 cm)	Water well and fertilize every two weeks when fruits start to swell. Pinch out growing tips when three leaves have formed. Pollinate as for summer squash. Put young fruits on wood or a slate to discourage insects and keep them clean	Cut pumpkins when they have fully ripened. This is usually in early fall, although small-fruited varieties will be ready in late summer. If there is any danger of frost, cover the plants with a row covers at night
Both types: Thin seedlings to 3 in (7.5 cm) when large enough to handle. Thin to final spacing one month later. (Thinnings may be eaten)	**Summer:** 1 ft x 1 ft (30 x 30 cm) **Winter:** 6 x 12 in (15 x 30 cm)	Hoe to control weeds **Summer spinach:** Water well. Will bolt in hot, dry conditions **Winter spinach:** Protect with row covers or straw from early winter	Pick leaves by pinching the base of the leaf stalk as soon as they are large enough. Take only half the leaves from a plant at one time, picking the largest first. Summer spinach can be picked harder than winter
Pull out weaker seedlings as they emerge. Harden off indoor-grown seedlings in late spring and plant into cropping site in early summer	**Small squash:** 3 x 3 ft (90 x 90 cm) **Medium squash:** 3 x 4 ft (90 x 120 cm) **Zucch ini:** 2 ft x 2 ft (60 x 60 cm)	Hoe until leaves smother weed growth. Water well, soaking the surrounding ground. Protect young plants from slugs. Pinch out growing tips of trailing marrows when they have four or five leaves to encourage sideshoots to grow. Pollinate by hand (*see* page 98)	Cut zucchini when 4 in (10 cm) long; keep cutting to encourage more to grow. Test by pressing a rib close to the stalk—it should yield slightly. Harvest squash when young, at the end of summer and cut them all by midfall, before any danger of frost

CULTIVATION TIPS

The charts on pages 94–97 give the basic cultivation advice for the vegetables covered here, but there are some additional tips that can help to ensure a good harvest.

Artichokes may be grown from seed, but it is quicker and easier to grow them from divisions, separated from existing plants, or bought through mail order. Jerusalem artichokes are grown from tubers. It is important to plant new tubers each year, as this reduces the chance of disease, and stops them taking over.

Celery that needs blanching is grown in deep trenches. Self-blanching celery is usually grown in a block to blanch the stems (of the inner plants, at least) naturally. You can also grow them in a cold frame with the top removed. The walls of the frame give shade to the plants, blanching their stems.

Corn is also grown in a block, to increase the chance of good pollination between the top flowers and those that form the cobs.

CULTIVATING CELERY

1 Before hilling up, wrap newspaper, cardboard, or black plastic around the stems of each plant to keep them free from dirt and to blanch, if necessary.

2 Plant out celery seedlings in trenches at the required distance and water well.

3 Begin the hilling up process, keeping the leaves exposed. Continue to water well.

4 Take care when harvesting celery not to damage the plants.

CULTIVATING SQUASH

PLANTING

Use a dibber to make holes in the soil, with a string line as a guide. Sow seeds individually.

PINCHING OUT

Pinch out the growing tips of squash to encourage the sideshoots to grow and the plant to spread.

PROTECTING

Place plants on pieces of wood, tile, or brick to protect the fruits from mud.

POLLINATING

Squash must be pollinated to bear fruit. There are two ways of transferring the pollen from the male to the female plants: either use a soft paintbrush dabbed inside each flower or insert the male flower itself into the center of the female (above).

Onions may be grown from seeds or sets of tiny, immature bulbs which will each produce one large onion. Sets are easier to grow, but keep them in a cool, dry place if you cannot plant them immediately, to discourage premature sprouting. Shallots are also grown from sets, but in this case each bulb multiplies to produce about six new ones. After harvesting, keep some of the shallots to plant for the following year's crop.

PESTS AND DISEASES

There are many potential pests and diseases in the garden; some of the most likely problems are listed here.

Leafhoppers and tarnished bugs can affect celery and celeriac. Protect plants with row covers. Affected plants will recover more quickly if they are given liquid fertilizer.

Dry growing conditions and poor soil can make artichokes small, woody, and unpalatable. Apply a general fertilizer to the soil and mulch with well-rotted organic matter to improve conditions.

Squash and pumpkins that are kept short of water, or spinach grown on poorly drained land, may suffer mildew on their leaves. If this is not severe, it will not affect the size of the crop, and infected leaves can be picked off spinach plants. Mildew is less likely if spinach plants are thinned early, as recommended. Squash can also suffer from viral diseases (*see* page 146).

Other potential problems to look out for on your crops include onion fly (page 148), nematodes (page 145), slugs (page 151), white rot (page 151), downy mildew (page 146), squash bugs (page 150), and celery leaf spot (page 146).

Growing Asparagus

Growing your own asparagus is an inexpensive way to put this often pricey ingredient on your plate. It does take a while to get established—three years if you start from seed—but the same plants will go on producing yields for up to 20 years.

Most hybrid asparagus are male plants; these produce more spears and do not produce seed. If you buy two-year-old crowns, you will be able to start cutting stems the second season after planting. Growing from seed will take a little longer. Asparagus does need a period of winter dormancy, so it does not do well in places with frost-free winters.

Asparagus doesn't have to be an expensive luxury. If you have space for a dedicated asparagus bed you can harvest your own.

Soil and situation

The most important requirement for asparagus is that the soil is well-drained. Ideal conditions are a deep, rich, light sandy loam. As the site is so permanent, it should be well prepared by digging in lots of organic matter the fall before planting and all perennial weeds must be removed, using a herbicide if necessary to insure that any deep roots or runners have been destroyed. If the soil is acidic, lime it to give a pH of 6.5–7.

When choosing where to put your asparagus bed, select an open sunny site, well sheltered from the wind. South-facing is usually best.

Sowing and planting

To raise asparagus from seed, sow it in outdoors in early spring. Sow in drills about 2 in (5 cm) deep and 1 ft (30 cm) apart. The seed is slow to germinate, so soak it in water for a day first to soften it. Water the emerging seedlings well through the summer and thin them to 6 in (15 cm) apart when the young plants are large enough to handle.

Asparagus is more typically grown from one- or two-year-old crowns. Two-year-old crowns may produce asparagus quicker, but one-year-old crowns are cheaper and less likely to suffer during transplanting. If you cannot plant them immediately, cover the roots in damp peat or damp burlap: they must not dry out.

To plant the crowns, dig a trench 8 in (20 cm) deep and 1 ft (30 cm) wide. If you have any fine gravel, scatter some at the base to improve the drainage. Put a little soil back into the trench to form a mound about 3 in (7.5 cm) high. Spread the crowns over this, so the roots are splayed out on either side of the ridge. Then cover them with about 3 in (7.5 cm) of soil. Leave 18 in (45 cm) between plants and 4 ft (1.2 m) between rows.

CULTIVATION TIPS

Good preparation and care when the plants are getting established will pay dividends in subsequent years with a prolific crop.

Fill in the trench gradually as the plants grow, always keeping the crowns just covered. As the foliage grows in spring and summer, you may find it better to support it than to let it grow really wild and straggly. Do not be tempted to cut back the foliage, as this will reduce the crop.

Keep the bed well-watered; it must never dry out. For the first year, weeds can be controlled by hoeing; after that they are best pulled out by hand, because hoeing could damage the plants' delicate roots.

After the first frost, in early fall, cut down the foliage to 3 in (7.5 cm). It should have turned yellow by now. Do this before any berries start dropping to the ground and pick up and destroy any that have already fallen; if they germinate, they will crowd the plants already established.

If you like blanched asparagus stalks, hill up the soil around the crops in fall and spring, making a mound of about 2 in (5 cm). Each spring, apply an all-purpose fertilizer to the soil, and each fall a mulch of well-rotted compost or manure.

CULTIVATING ASPARAGUS

1 Plant crowns on top of mounds dug in trenches and carefully cover with soil.

2 For the first year, do not cut the foliage, but continue to water the plants well. Control weeds by hoeing.

3 After the first frost, cut down the foliage before any berries begin dropping to the ground.

4 In the following spring, apply fertilizer and a mulch of compost or farmyard manure in the fall.

5 If growing two-year-old crowns, cut a few spears the following spring, but do not harvest too many.

6 Established plants may need to be supported with stakes, but take care not to damage the crowns.

Harvesting

If you planted one-year crowns, do not cut any spears the first spring (one year after planting). Two-year-old crowns will give a small harvest the following year, but pick this sparingly—no more than one spear from each crown. When the plants are four years old, you can harvest the spears for four weeks, and in following years, for six weeks. The plants are ready for harvesting when the spears start to form.

Cut the spears using a special curved knife, about 3 in (7.5 cm) below the surface of the soil. The spears should have about 5 in (12.5 cm) showing above the ground if they are to be tender and succulent. Any taller than this, and they will be woody and stringy to eat. Always stop cutting in late spring to allow the plants to build up their resources again before the next growing season. Contain the summer foliage if it gets unruly and cut it down each fall as described above.

PESTS AND DISEASES

The most common problems to look for in asparagus beds are black slugs (page 151), asparagus beetle (page 144), violet root rot (page 148), and rust (page 146).

Watch out for changes in the foliage that can signal an attack: distorted, patchy brown leaves indicate asparagus beetle, reddish raised areas are rust, while yellowing foliage can be a sign of violet root rot.

Greenhouse Vegetables

A greenhouse is not an essential item of equipment for those who want to grow their own garden produce. But in short-summer climates, a greenhouse can boost the yield and the quality of some crops, speed up the time between sowing and maturing, and broaden your range of possible crops for growing.

Before you buy a greenhouse, weigh up the cost plus the maintenance of heating it (if you choose to do so) against the value of the crops you will produce. You may feel it is worth it even if the math does not agree, but it could be many years before a harvest of peppers will pay you back. For more information on choosing and looking after a greenhouse *see* pages 34–38.

EGGPLANT

This Mediterranean plant can be grown outdoors in many climate zones, but you will get a better crop in cold areas if you grow eggplants in a cold frame, in a greenhouse, or even in a sun room.

Sow seeds in early spring and put them in a heated propagation to germinate; if you don't have a propagation unit, sow in spring and put them on the side of the greenhouse that gets the most sun. When the seedlings are large enough to handle, prick them out into small pots filled with soilless potting mix. When they are 6 in (15 cm) high, put them into their permanent site—large containers, raised beds, or straight into the greenhouse bed. Space plants 18 in (45 cm) apart.

You may be surprised how easy it is to grow heat-loving crops like eggplants, chili peppers, and cucumber in a greenhouse.

When the plants are 9 in (25 cm) high, pinch out the growing tips to encourage side growth. As the fruits form, pinch out the tips of the branches so that there are no more than two or three fruits on each branch. Remove any sideshoots. As the fruits begin to swell and become heavy, you may need to support the plant with stakes to stop it drooping.

Water the plants well—they dry out easily. The fruits will be better if you feed the plants each week with a liquid fertilizer. Spritz the leaves with water in warm weather as this discourages red spider mites—the most virulent pest in the greenhouse.

Pick the fruits when they are swollen and a shiny, deep purple color (some varieties are mottled or have creamy white stripes, too). Cut them free of the plant with scissors or a sharp knife, handling them carefully to avoid bruising.

CUCUMBERS

Instructions for growing outdoor cucumbers are given on pages 82–86, but they need a long, warm growing season not found in northern gardens. The fruit of greenhouse plants will also be larger and have less tough skins.

If cucumbers develop from male flowers, they will be bitter, so either remove male flowers as they appear or choose self-pollinating varieties that produce only female flowers. Tiny cucumbers form behind the female flowers, while the male flowers just have a simple stem, making them easy to distinguish.

Raise seed in late winter in a propagation unit and pot on in the same way as for eggplant seedlings. Plant them in their permanent homes when the seedlings have begun to develop their rough leaves.

Grow bags and rings

Grow bags are very useful in a greenhouse. Raise the plants in the usual way and transplant them into the grow bag, with its rich mixture of potting soil, fertilizer, and water-retaining substances.

An alternate method of growing greenhouse tomatoes is in a ring culture system (*see* page 89). You can use rings that fit into a grow bag, cylinders of hardware cloth, or bottomless pots or rings that are filled with potting soil and placed on a bed of gravel or crushed stones. The roots grow into the gravel searching for water, while fertilizer is applied to the potting soil in the ring for the "feeding roots."

Staking and support

As the plants grow, you will need to support them. This can be done in one of several ways. When the plant is about 9 in (25 cm) high, tie a string very loosely a round its base and take this up to the top of the wall of the greenhouse, attaching it to another string running up the roof.

As the plant grows, pinch out the sideshoots and train the main stem up the string by twisting it gently. To do this, the string should not be too tight, and the plant will need very careful handling to ensure you do not break the stem. When it reaches the top of the wall of the greenhouse, you can continue taking it up the string that runs parallel with the sloping roof.

The other method is to put a stake into the ground beside the plant and let it grow level with the top of this, tying the main stem loosely to the stake to support it. Fix horizontal wires or strings at intervals up the stakes, then pinch out the growing tip of the cucumber when it reaches the top of the stake. As the laterals grow, train them along the strings, twisting them carefully around them.

Routine care

Keep plants well watered, and spray them if it is very hot. You will get better cucumbers if you feed the plant with liquid fertilizer when the fruits are beginning to swell, but this is only necessary every two weeks.

Cucumbers will be ready for harvesting from about midsummer, and it is best to pick them as soon as they are ready. This is when they have nice straight sides, not when they have reached their maximum size. Picking cucumbers at this stage will also encourage the development of more fruits.

Hints on growing greenhouse crops

- Sow the seeds thinly in seed trays or cell packs, pricking out later into individual pots.
- Most greenhouse crops will need to be trained up some form of support to prevent the plants getting straggly.
- The growing tips of many greenhouse crops, tomatoes especially, should be pinched out.
- Continue to spray the plants with water to increase humidity and discourage pests. Feed with liquid fertilizer.
- As soon as any of the crops in the greenhouse are ready, harvest them to make room for the remaining crops.
- Do not leave crops until they have reached their maximum size before picking. Always cut them with a knife.

TOMATOES

Many gardeners can grow tomatoes outdoors (*see* pages 87–90), but you will get bigger yields in short-summer regions by growing them inside a greenhouse.

Tomato seedlings are raised from seed in the same way as the other crops so far discussed and the seed should be sown in early spring. They can be set in their permanent site when they are about 4 in (10 cm) high, with about 18 in (45 cm) between the plants and 2 ft (60 cm) between the rows. Stagger the plants in the rows so they are not directly in line with one another.

The easiest and most efficient way of supporting tomatoes growing in a greenhouse is to tie strings loosely round their bases and take these up to a horizontal string hung across the greenhouse, level with the top of the walls. As the plants grow, gently twist the main stems around the strings, taking great care not to snap the stems. Nip out the sideshoots that emerge from the leaf axils (*see* page 90) and pinch out the growing tips when the plants near the top of the horizontal strings.

Feeding and watering

Mulch the tomato plants when you plant them and then water them once a week. Even, regular watering is important with tomato plants; if you allow them to dry out and then give them a healthy drink to compensate, you are likely to get split fruit. Feed with liquid fertilizer every 10 days or so when the tomatoes have begun to develop.

As the plant grows, the leaves at the base will begin to turn yellow and die back. At this point it is best to pick them off. This makes it easier to pick the fruit and water the plants.

Take care not to strip off too much foliage, or you will leave the plant with no means of manufacturing food for the growing tomatoes.

Harvesting and ripening
You can pick tomatoes as they ripen, or you can take them green from the plant and ripen them on windowsills or in other warm, sunny places. Tomatoes that are still green at the end of the season can often be ripened successfully in the fruit bowl, with some bananas. Harvest the tomatoes by supporting them in your hand and nipping off the stem just above the calyx.

PEPPERS

Warm greenhouse conditions are ideal for these hot-weather crops. Bell peppers will be sweeter and chili peppers more fiery when the plants are grown in a greenhouse.

Round, sweet, bell peppers and the thinner, longer and very much hotter chili peppers can be grown successfully in a greenhouse and are treated in the same way as these other greenhouse crops.

Sowing and care
Sow the seeds in spring and raise the seedlings in a heated propagation unit. After that, treat the young seedlings exactly the same as those of eggplants and cucumbers, planting them in their permanent site when they are about 4 in (10 cm) high. If growing directly in soil, put them about 18 in (45 cm) apart and push a heavy stake into the soil beside each plant.

Spicy chili peppers will thrive in a greenhouse; if you don't have one, try growing them on a sunny windowsill or sunroom.

The fruits can be very heavy and plants will need support to prevent them drooping as the fruits develop.

These plants need very little attention; just water them regularly and feed them with liquid fertilizer every week or so when the fruits begin to appear. Support them by tying them loosely to the cane as this becomes necessary.

You can either pick bell peppers when they are green (in midsummer) or you can leave them on the plant to ripen still further. They will either turn red or yellow, depending on the variety. Cut the fruit off the plant using scissors or a sharp knife in the same way as eggplants.

GREENHOUSE MANAGEMENT

The conditions in a greenhouse that are so good for the crops are also ideal breeding grounds for pests and disease. So pay extra attention to the crops you have and nip any problems in the bud before they get beyond your control.

Crops may be grown in the greenhouse in the greenhouse beds, in large pots or in fertilized grow bags. Tomatoes may be grown in a special system known as ring culture (*see* page 89), in which the plants are actually grown in bottomless pots filled with potting soil and set on a bed of gravel or sterile aggregate or into a growing bag.

The easiest method is to grow in the greenhouse beds, but if you mainly use the greenhouse for growing tomatoes, you must either change the soil each year or sterilize it before planting again. If you grow tomatoes in the same soil in the greenhouse for two years running, they are prone to various diseases.

You can follow tomatoes with a different crop, such as peppers or cucumbers, which will not be diseased in the same way, but be prepared for tomato seedlings to sprout alongside. The soil should be changed after these crops have been harvested, and tomatoes can be grown in the site the following year.

Changing the soil

In early summer, dig out and swap the soil in the greenhouse beds for some in the garden. Mix in plenty of well-rotted compost or manure at the same time, and you should have even better tomatoes the following year. You can put the old soil back into the greenhouse after a year, when it has been thoroughly exposed to the elements and has had a chance to be washed through by the rain. To sterilize soil, spread it out and cover with sheets of plastic, leaving it to "cook" for several weeks in the sun.

Avoiding pests and disease

Always keep the inside of the greenhouse clean and tidy (*see* page 38); if weeds do appear, pull them out right away and check over all the crops each day. Regular watering will not just prevent plants from drying out, but can also discourage pests, which often thrive in hot and dry conditions.

Growing Herbs

Although growing herbs may not always represent the same saving in cost as vegetables, they are so important in adding interest and flavor to your cooking that it is well worth producing a selection of the ones you like to use the most.

Herbs are easy to grow. Although they mostly flourish best in rich, well-drained soils and sunny, sheltered positions, they will generally grow in almost any type of soil or garden site. In addition, of course, individual herbs take up only a small amount of space and they can be grown in any convenient spot, even in pots on a windowsill, deck, or patio.

Herbs are ideally sited close to the kitchen, so they are handy when you need them. If this is not possible, they can be grown at one end of the vegetable garden, or scattered individually around the flower beds. Many are decorative enough to grow in a border or bed and, in addition to adding color, they will also help pollination among the flowers by attracting bees.

Many herbs can be preserved for out-of-season use, either by drying, making into jellies, freezing, or making herb-infused oils and vinegars. See the chapter on Preserving your Produce (pages 204–36) for more advice.

Planning a herb bed

A formally planned and well laid out herb garden was a traditional feature of many colonial gardens. Lack of space has made them less common, but if there is room, they can become a charming and productive addition to your kitchen garden.

TAKING HERB CUTTINGS

Many new herb plants can be raised by taking cuttings from existing plants or by digging up part of the plant or bush and dividing the roots.

1 Take heel cuttings by removing sideshoots from the main stem.

2 Dip the cut end in hormone rooting powder then push into pots of potting soil. Cover the pots with miniature cloches or a clear plastic food bag or stand them in a propagation tray until the cuttings are well established.

3 Dividing the roots gives you more established plants, but fewer new ones. Dig up the existing plant and carefully tease the rootball apart, taking care not to damage the roots. Disentangle the stems and replant the original plant and the new ones immediately.

Traditionally, herb gardens followed geometric shapes—triangles, circles, squares, and so on—in the style of an Elizabethan knot garden. Each small area would be bordered by a neatly clipped edging of boxwood (*Buxus*) or other small-leaved evergreen. Creating an elaborate pattern like this is a long-term plan, waiting for the herbs to establish and the evergreens to grow, but you can use a wooden framework to divide a bed or create a patchwork of different herbs that does not need a formal edging.

If you are planning a herb garden, bear in mind that the taller herbs should be placed at the back so they do not put others in perpetual shade. Some herbs like partial shade and will grow happily when planted close to their tall companions.

Where space is at a premium, herbs can be grown successfully in tubs, pots, or other containers, placed anywhere that there is room, from patios and decks to kitchen windowsills or the side of a flight of outdoor stairs. Some herbs will even grow in narrow gaps between paving stones in a path or terrace, but not in heavy traffic areas, where they will be repeatedly trampled.

Choosing what to grow

The choice of herbs to grow is likely to depend on individual taste, although those most widely used in cooking are probably chives, basil, parsley, thyme, rosemary, and sage. Try some of the more unusual ones as well though—they are usually no more difficult to grow and can be much more interesting in the kitchen.

Many herbs are perennial, although in some instances it is best to treat them as annuals and start them again each year to ensure that you have a good supply of fresh young leaves and stems, rather than woody growth. Few will withstand very heavy winter picking, so allow any you wish to keep for next year to rest over winter.

Herb plants are pretty as well as productive—and a useful ally in the fight against pests. Thyme and lavender will give off a delicious scent and attract insects that help to control aphids.

CULTIVATION

Growing advice for a variety of common herbs is given on pages 106–07. As most herbs do best in similar soils, they do well planted together in a dedicated herb bed.

The spacings given on the next page refer to space between plants; in most home gardens, you will not need to plant more than one row of any particular herb; in fact, one plant of each herb will provide sufficient leaves or seeds for most families.

Parsley and basil are the most notable exceptions to this rule, as they are so widely used in the kitchen and are often used in large quantities when you need them. Many households find they need several plants to fulfil their regular needs.

Sow herb seeds very sparingly, and thin the seedlings progressively, discarding the weakest ones until you are left with the number you need. If sowing indoors, sow two or three seeds to a pot, and discard the weaker seedlings as they emerge.

Pests and diseases

As a rule, herbs do not suffer too much from pests and diseases, and these will be kept to a minimum if normal garden hygiene is observed.

When herb plants are affected, it is probably best to dig up the affected plants and discard them, planting fresh seeds or plants to replace the lost ones.

CULTIVATION OF HERBS

HERB	SOIL/POSITION	SOWING	SPACING
Balm (evergreen perennial)	**Soil:** Any **Position:** Sunny, but will tolerate some shade	Indoors in pots or outdoors in late spring, in ½ in (1 cm) drills	1 ft 6 in (45 cm)
Basil (annual)	**Soil:** Any that is well-drained **Position:** Sunny and sheltered. Bush basil may be grown in pots indoors for winter	Indoors in pots or outdoors in late spring, in ¼ in (0.5 cm) drills	1 ft (30 cm)
Bay (evergreen perennial)	**Soil:** Any **Position:** Sunny and sheltered. Grows very successfully in a large barrel or pot	Buy a young plant or take 6 in (15 cm) cuttings in spring and grow in potting soil until established	One tree is sufficient
Borage (annual)	**Soil:** Any that is well-drained **Position:** Sunny	Directly outdoors in midspring and again in late summer. Sow in ½ in (1 cm) drills	1 ft (30 cm)
Chervil (biennial)	**Soil:** Any that is moist and water-retentive **Position:** Sheltered, in partial sun	Directly outdoors at regular intervals from early spring through summer in ¼ in (0.5 cm) drills	1 ft (30 cm)
Chives (perennial)	**Soil:** Any that is moist **Position:** Sunny or partial shade	If raising from seed sow directly outdoors in ¼ in (0.5 cm) drills in midspring	1 ft (30 cm)
Coriander (annual) (also called cilantro)	**Soil:** Rich and well-drained **Position:** Sunny; near dill and chervil	Directly into cropping site in spring and late summer, in ¼ in (0.5 cm) drills	1 ft (30 cm)
Dill (annual)	**Soil:** Well-drained **Position:** Open and sunny	Directly outdoors in midspring ½ in (1 cm) drills	10 in (25 cm)
Fennel (perennial)	**Soil:** Rich and well-drained. Will grow on well-cultivated soils **Position:** Warm and sunny. Does not flourish well if grown near coriander	If raising from seed, sow at regular intervals from mid- to late spring in ½ in (1 cm) drills. New plants may also be obtained by division	1 ft (30 cm)
Horseradish (treat as an annual)	**Soil:** Deeply dug light loam that is well-drained **Position:** Sunny or partial shade	Plant roots or root cuttings 5 in (12.5 cm) long in March so they are 2 in (5 cm) below the ground	1 ft 6 in (45 cm)
Marjoram or Oregano (treat as an annual)	**Soil:** Light and well-drained **Position:** Sunny but sheltered	Directly outdoors in spring and/or early fall, or raise from cuttings	10 in (25 cm)
Mint (perennial)	**Soil:** Most, but likes rich, moist soil best **Position:** Full sun gives best flavor	Plant young plants in early spring in a bucket or similar container to contain the invasive roots. Hard to grow from seed	6 in (15 cm)
Parsley (treat as an annual)	**Soil:** Rich and well-drained, but moist **Position:** Sunny	Sow outdoors at intervals through spring and early summer, in ¼ in (0.5 cm) drills. Pour boiling water along the drills to aid germination. Raise indoors in pots	8 in (20 cm)
Rosemary (evergreen perennial)	**Soil:** Light or sandy. Well-drained **Position:** Sunny but sheltered. Near sage	Sow into growing site or seed bed in April, or raise from heel cuttings taken early summer	
Sage (biennial)	**Soil:** Well-drained **Position:** Sunny, but sheltered	Sow late spring to summer in ¼ in (0.5 cm) drills, or take heel cuttings in summer	Two plants 1 ft (30 cm) apart
Tarragon (perennial)	**Soil:** Well-drained **Position:** Sunny and warm	Buy a young plant and plant in early spring or fall, or raise from cuttings taken in spring or fall, or from root division	Only one plant is needed
Thyme (perennial) The different types are all cultivated in the same way	**Soil:** Light, sandy, but well-drained **Position:** Sunny	Sow outdoors in spring and summer in ¼ in (0.5 cm) drills or indoors in trays. Raise new plants from heel cuttings taken in May/June or by root division	9–12 in (25–30 cm)

THINNING/TRANSPLANTING	CULTIVATION	HARVESTING
Plant out indoor plants in early fall. Thin sowings in cropping site when large enough to handle	Water well in dry spells. The year after planting, cut back plants in early summer. In fall, cut to just above ground level	Pick very sparingly the first summer. After that, pick as required throughout late spring and summer. Pick leaves for drying in early summer
Harden off plants in spring and plant out in early summer	Water well in dry spells. Pinch out growing centers to encourage bushy, leafy growth and pinch out flowers as they form	Pick leaves as required through the summer. The first frost will kill outdoor plants
Thin sowings when large enough to handle	Bay can be trimmed into a variety of shapes in summer. Trees grown in containers are best put in a light, airy shed in the winter	Pick leaves as required. For drying, cut branches on summer mornings
Thin when seedlings are large enough to handle	Keep weed growth in check. Borage grows very quickly and reseeds itself readily	Pick as required. For drying, pick undamaged leaves in the morning after the dew has evaporated
Thin to recommended spacing as seedlings emerge	Water well in dry spells and pick off flowering stems as they appear	Pick as required. For drying, pick sprays from mature plants on a sunny morning
Thin to recommended spacing as seedlings emerge	Water well in dry spells and pick off flower heads to help prevent early dying back. Dig up and divide in spring or fall	Pick as required, cutting leaves at the base. Picking encourages heavier growth
Thin when seedlings are large enough to handle	Hoe to control weeds and water in dry spells	Gather seeds by picking flowerheads when they smell spicy and seeds have turned from green to beige. Dry heads in boxes, trays, or racks, then shake out seeds
Thin seedlings when large enough to handle. Leaves from thinnings may be used	Hoe to control weeds and water in dry spells. Stake plants as they grow	Pick leaves as required before flowers appear, cutting to the base to encourage growth. Leave some plants to flower and produce seeds
Thin to recommended spacing when large enough to handle	Needs no special attention	*See* Dill
Plant no more than three or four, because they spread	They need no special attention except for watering in dry spells. Dig up all plants each year and replant new ones	Dig up roots from late summer through fall. Lift all plants and store roots as described on page 207
Thin to recommended spacing when large enough to handle	Need no special attention	Pick leaves as required before flowers form through summer and fall. Leave the young central leaves
	Water well after planting. Dig up some roots in fall and grow in containers over winter. Cut plants to ground in midfall	Pick as required through spring and summer. Young leaves have the most flavor. For drying, cut stalks as the first flowerheads appear
Thin seedlings twice to achieve recommended spacings	Water well in dry spells and protect plants with cloches in winter if trying to keep them until following spring	Cut out flowering stems to encourage leaf growth. Pick leaves sparingly until plants are established, then pick them hard, leaving only the green center. For drying, cut from mid-summer onwards
Gradually discard seedlings until one healthy plant is left	No special attention	Pick fresh as required (do not pick too heavily during winter). For drying, pick as plants come into flower
Thin seedlings as they emerge and transplant in fall	Pinch out flowerheads and growing tips to encourage bushy growth. Cut back hard after flowering	*See* Rosemary. The flavor is at its best in mid-summer
	Water well in dry weather. Plant dies down in winter: protect in cold areas with straw	Pick fresh leaves as required through summer and fall. For drying, pick leafy sprays from fully grown plants in the morning
Thin to recommended spacing. Prick out indoor sowings to individual pots and plant out in fall when well-established	Water growing plants well and cut back after flowering. Pinch out growing shoots to encourage more leaf production and bushy growth	As for Rosemary. Do not cut too heavily in the fall or the plant may not survive the winter

GROWING FRUIT

Growing fruit

By and large, fruit falls into two categories—tree fruit and soft fruit. Many older city and town gardens contain at least one fruit tree. You don't need to plant an orchard: almost all gardens will have room to grow some fruit and with more than a couple of trees you can easily find yourself with more fruit than you can possibly eat, so you can preserve it for future use.

Soft fruits are the taste of summer. Most yards have room for a wide variety, including currants, strawberries, gooseberries, raspberries, and blackberries.

Fruit is generally seen as less essential than vegetables in a self-sufficient garden, but a couple of fruit trees and a handful of bushes will produce a surprisingly large quantity of fruit that will make a significant contribution to your family's table. Most types of fruit are easy to freeze and to preserve in other ways that will see you through the winter.

A row of raspberry canes can be used as a productive windbreak to shelter more tender plants behind, and an apple tree makes a charming addition to the ornamental garden. To save space, many tree fruits or vines can be trained against a wall as espaliers or in a fan shape. They are not hard to incorporate into an established vegetable garden or to fit into a new kitchen garden plan.

Fruits in two main groups

The tree fruits covered in this book include apples, pears, peaches, nectarines, plums, damsons, gages, apricots, cherries, figs, and citrus fruits. The soft fruits include raspberries, loganberries, blackberries, blueberries, currants, gooseberries, and strawberries.

Grapes, kiwi, melons, and rhubarb do not fit into either category—grapes and kiwi grow on perennial vines, melons are annual plants, and rhubarb grows from tubers, but they are all popular fruits and you will find all the growing advice you need in these pages.

Commercial vineyards can be found as far north as Ontario and Minnesota, but for edible grapes in short-summer zones, you will need a warm microclimate, a south-facing wall, or a greenhouse. Melons also need plenty of warmth, but watermelons are too sprawling for most greenhouses. Smaller cantaloupes and honeydews can fit into a small greenhouse, however.

Modern and compact trees

Tree fruit grown on traditional standard or half-standard trees can reach heights and spreads of 20 ft (6 m) and more, which can make picking difficult and possibly dangerous. Even if a garden is big enough to take a couple of these trees, it is rarely an efficient allocation of space, which would be far more productively used growing other fruit and vegetables or supporting livestock.

Most tree fruit bred for planting in home gardens now is produced by grafting the different varieties onto special "dwarfing" rootstocks. This restricts the tree's growth in order to keep it manageable, but also gives quicker fruit production. Where once it took 10 years or so for trees to start to bear fruit, most modern trees will now do so in their third year.

Make sure when purchasing tree fruit that the trees you buy are growing on the right size rootstocks, which are numbered according to their size. Check the plant label for a guide to the ultimate size of the tree in maturity.

Keeping trees healthy

To make sure that your fruit trees remain healthy and productive, they must be pruned regularly. This will help to minimize the risk of disease, which is greater when branches are crowded or crossing within the tree, but will also concentrate the tree's energy into producing fruit, rather than unnecessary branches and foliage. Left to their own devices, trees will continue to produce an abundance of fruit, but the quality and taste will almost certainly deteriorate with each year.

Sweet and simple soft fruit

Soft fruit has no such size problems. The different types grow either on bushes (blueberries, currants, and gooseberries) or canes (raspberries, loganberries, and blackberries). Although these also need regular pruning and training if they are to be fully productive, this is a less complicated task than for tree fruit.

Strawberries grow on neither bushes nor canes, but on low, spreading plants, which put out runners that produce new plants as they grow. These new plants can be separated from the parent plant and transplanted elsewhere or passed on to friends. Save some to renew your plants every couple of years.

Choosing what to grow

There are endless varieties of all types of tree and soft fruit. Some trees need a period of winter dormancy; others must have hot weather to thrive. The best way to choose what to grow in your particular garden is to consult a local nursery or your local Cooperative Extension Office to see what grows best in your area. Many nurseries stock a limited number of varieties, because over the years they have established which types are the most successful for their customers. If you wish to try growing more exotic varieties, there are a number of high-quality mail-order nurseries that sell dwarf fruit trees and cane fruit.

Planting Out

Most fruitS are bought as bare-root trees, bushes, or canes, rather than being grown from seed. They should be planted during winter, when they are dormant, but not when the ground is frozen or waterlogged. The ideal time is in early and mid-winter, but planting can be done any time through to the following spring. It is best to buy two- or three-year-old fruit trees and soft fruit bushes, and one-year-old fruit canes.

If you will not be able to plant the fruit for a few days after buying it, leave the roots encased in a protective wrapping and keep the plant in a cool shed. The day before planting, inspect the roots and if they are dry, soak them in a bucket of water for a few hours.

Planting a bush

Dig a hole wide enough to take the roots of the bush and fork well-rotted organic matter into the bottom. Place the bush in the hole, spread out the roots, and cut off any that are growing upwards.

Fill in around the roots with soil and firm it down. When planted, the base of the bush should be just covered with soil. Mulch the base with compost, straw, or leafmold.

PLANTING A FRUIT TREE

1 If the roots of the tree are dry, soak them in water before planting. Place a layer of compost in a wide hole, position the tree, and spread out the roots.

2 Mix soil with compost and fill in around the roots. Shake the tree to settle the soil. Firm gently with your feet, top up if necessary; firm again.

3 Water around the roots well and give the soil a mulch of garden compost, bark mulch, straw, or leafmold. Secure the top of the stem to a supporting stake with a tree tie or adjustable strap.

Planting canes

Raspberry and other fruit canes look like unpromising sticks when you buy them, but they will bush out.

Dig well-rotted organic matter into the bottom of a trench. Position the canes, spreading out the roots and keeping the canes upright. Pack around the roots with soil to firm the cane into the ground, then finish replacing the remaining soil dug out of the trench and firm it down again. Cut back the canes to around 9 in (25 cm) high.

Apply a mulch of compost, straw, or leafmold along the row.

Cut back canes to just 9 in (25 cm) after planting to encourage vigorous new growth.

Space canes 18 in (45 cm) apart along a trench that is around 3 in (7.5 cm) deep. If you are planting more than one row, space them 6 ft (1.8 m) apart.

Protecting plants

Because the best time for planting trees, bushes, and canes is in the winter, the plants may need to be protected against the harsh weather, or animals in search of food while it is scarce elsewhere. Windbreaks will block winter blasts; cages may be needed to guard against deer, birds, and rabbits. Soft fruit may need to be protected from animals by being grown in cages of hardware cloth or chicken wire (below).

Supporting canes

Drive stakes into both ends of the trench where you have planted raspberry or other soft fruit canes. Stretch three evenly spaced strands of wire between them and train shoots to these wires as they grow.

Heeling in

Do not plant a tree until the weather conditions are right. If they are not, keep it temporarily in the shed; if it is a bare-root tree, keep the roots in damp straw until you are ready to plant it. A better option for a bare-root tree is to heel it into the soil in a sheltered part of the garden. Do this by planting it in the soil at an angle (left), covering the rootball with soil to help it to retain moisture.

Growing Apples

Just about every garden will have room for an apple tree or two. Early-maturing varieties will be ready for picking in late summer and the late varieties in midfall, but they will keep well for eating through the winter until midspring if you are careful as you pick, wrap, and store them.

Apples are not self-pollinating, so it is essential to grow at least two trees (that blossom at the same time) unless you have a neighbor who has a number of apple trees. An alternative, if you really feel you only have room for one tree, is to grow a multi-variety tree, which has two or more cross-pollinating varieties grafted onto a single rootstock.

There are myriad varieties of apples, divided into two basic types: cooking and dessert apples. Before buying trees, decide which type you want to grow. Choose varieties according to your taste, their cross-pollination compatibility, and what is known to grow well in your area. Avoid Cox's Orange Pippins where the soil is poor or if you live in a cold area, for example. This variety was developed in England and will not withstand harsh winters. Late-flowering varieties are best for cold-winter gardens.

How many trees do you need

The four different forms of apple tree (*see* pages 116–17) give the following approximate yields when they are at full capacity.

- **Cordon** Up to 10 lbs (4.5 kg), but an average of 4 lbs (1.8 kg)
- **Open center dwarf tree with four main branches** Up to 60 lbs (27 kg), with an average yield of 30 lbs (13.5 kg)
- **Espalier** The yield depends on the size of the tree. Expect about half the amount of a dwarf tree grown on the same rootstock.
- **Dwarf pyramid** 30 lbs (13.5 kg)

Soil and position

Apples like deep, fertile loam, but will grow in most well-cultivated soils providing they are not waterlogged or too acidic. Choose an open sunny position that is well-sheltered from prevailing cold winds and avoid planting in pockets or hollows where frost collects and lingers.

Planting and routine cultivation

Make sure the ground is completely free of weeds (especially perennial ones) before planting and keep the area around a newly planted tree weed-free, especially through the first year. Remove weeds with gentle hoeing or by hand, taking care not to damage the young tree roots.

Water new trees very heavily in the first growing season, and from then on water all trees in dry spells—trees are good at finding and taking up water, but will still be weakened in periods of drought and this can affect the blossom and, later, the quality and quantity of the fruit. Mulch the ground around the tree in spring until it is well established.

When you buy apple trees, ask the nursery for advice about how far apart to plant them. This differs so much with the variety, the form in which the tree is to be grown, the type of soil, and the

For the best and sweetest fruit, thin out clusters of apples in the early stages of their growth to just one or two fruits.

SEASONAL GROWING GUIDE

Late fall to early spring	**Winter pruning** Cut back hard to invigorate new growth
Early summer	**Thin out fruit** Pick off developing fruits to leave no more than two apples in each cluster
Midsummer	**Summer pruning** Cut out new growth to improve shape and avoid overcrowding
Late summer to late fall	**Harvest fruit** Pick fruits as they ripen. To store for later use, individually wrap unblemished fruits only in paper

rootstock on which it is grafted, that it is not practical to give specific measurements here.

TRAINING AND PRUNING

The eventual shape of any tree fruit is established by the pruning and training done in the first four years of its life. If you buy two- or three-year-old trees, the training will have already begun at the nursery.

The aim in pruning is to maintain the shape you have established, to allow a good penetration of light to reach all parts of the tree, and to maintain a constant good balance between growth and fruit. Enough new wood must be allowed to develop to turn into next year's fruiting growth and old wood should be cut out after its second or third season.

Be bold, and prune hard

Most people do not prune established trees hard enough—open center bush trees and dwarf pyramid trees in particular should be reduced by a third each winter. Winter pruning is invigorating—the harder you prune, the harder the tree will grow. Summer pruning is restricting as you are cutting out new growth as it forms.

SPUR- OR TIP-BEARING

Apple trees are either spur- or tip-bearing—depending on the variety, the apples are either born on the spur wood that grows from the leaders and laterals, or on the end of the previous season's shoots.

When pruning spur-bearing varieties, the leaders should be cut back by half each year and all the laterals cut back to three or four growth buds to induce the spurs to grow.

An alternative method to this is to operate a three-year cycle, which works on the principle that the second and third year wood produces the best fruit. For this you need a framework of up to six strong branches growing off a central trunk. The laterals, which will bear fruit in their second and third years, grow from these permanent branches. After their third year, cut out the entire lateral, so that a new one will grow in its place.

Tip-bearing tips

In tip-bearing varieties, it is obviously important not to cut off the shoots bearing the terminal buds, as it is these that will form the following year's fruit. The difficulty is how to keep the tree bushy and compact without cutting out the fruiting wood.

This is best done by sawing out entire big, old branches when they have ceased to be productive. This will shape up the tree and allow new branches to grow in their place. Do not just snip at the tree, as a light trim will have little effect—make big cuts that will count.

CULTIVATION TIPS

Apples, like other fruit trees, will produce fruit year after year without your help, but with some care and attention you can increase the quality and quantity of your crop.

If large clusters of apples are produced after a bumper blossom season, these should be thinned. It may seem contrary to remove fruit from the tree, but if you leave it all to mature, the results will probably be small and sour and prone to

GOLDEN RULES FOR PRUNING

A few simple rules will help to ensure that your pruning improves the health and vigor of your trees, rather than introducing disease or damaging the plant.

- Always make a positive cut with sharp, clean pruners or loppers. Do not tear the wood.
- Seal any cuts you make with a pruning seal to prevent disease entering.
- Always cut back to either a new growth bud or flush with the branch.

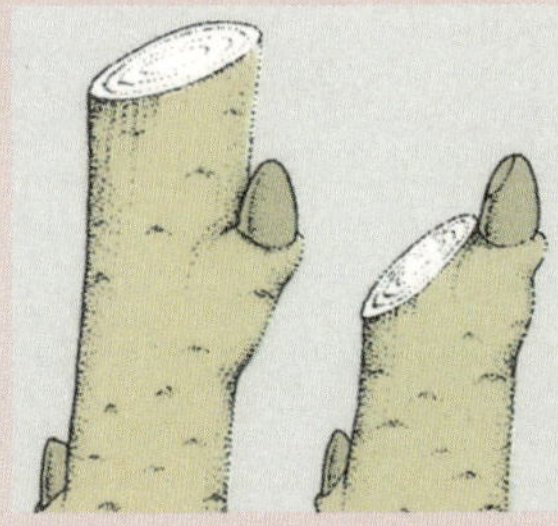

- Never cut between buds or cut so close that they might be damaged.

PRUNING TERMS

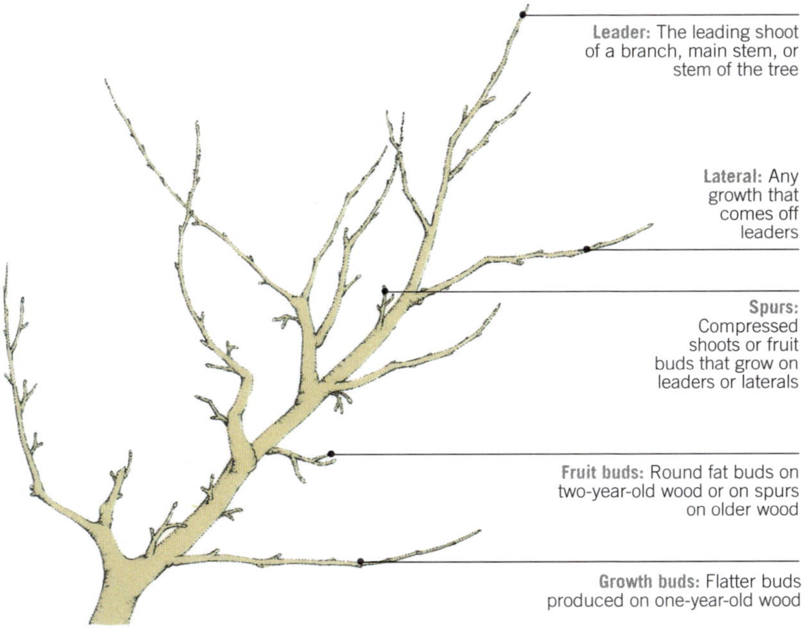

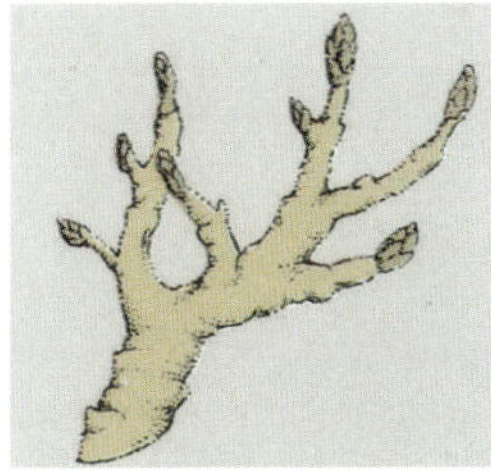

Spur

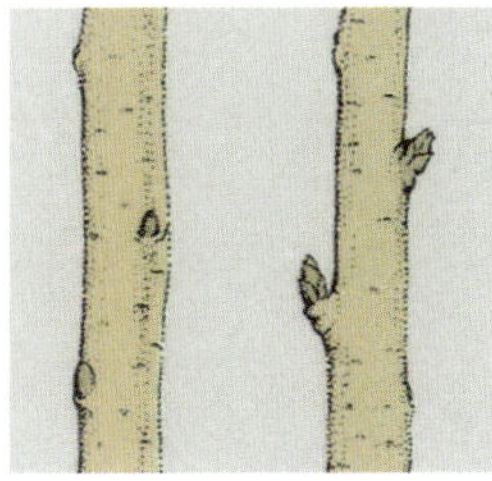

Growth and fruit bud

disease due to overcrowding on the tree. Thinned out, you will get fewer apples in all, but they will be far better tasting.

Trees will naturally thin their own crop in early summer—a process known as the June drop—but check the tree after this and cut out (with scissors) any damaged or misshapen apples that remain to leave one, or at the most two, to each cluster. Ideally, you should aim to keep the apples about 6 in (15 cm) apart.

Harvesting and storing

Test fruit for ripeness by supporting the apple in your hand and gently lifting it. If it is ripe it will come away easily from the tree.

Always handle apples very gently, particularly if you want to store them for eating later in the year, as any blemishes will spoil the fruit and this can spread to the whole box in storage. Pick apples with the palm of your hand, not your fingertips, which can bruise the fruit.

Some varieties of apple can be kept for many months, provided they are stored correctly. The most important rule is to only use perfect fruit and to store it in such a way that no apple touches another, by wrapping them individually in tissue or newspaper, for example. For preserving apples, *see* page 222.

PESTS AND DISEASES

There is a frighteningly long list of pests and diseases that can affect apple trees, but the main ones to fear are apple mildew (page 149), apple scab (page 149), and codling moths (page 149).

Old-fashioned protection practices, such as applying sticky tree bands around the trunk of the tree in summer and early fall, can help trap and therefore deter pests such as the codling moth, which climbs up the trunk of the tree to lay its eggs.

Some gardeners still recommend regular spraying with various pesticides through the spring and summer, but in most home gardens (particularly if strict garden hygiene is observed) this should not really be necessary.

If your trees are showing signs of disease, consult your local Cooperative Extension Office. Otherwise, ask advice from within your local gardening club or from neighbors with trees. Remember that over-spraying can do the tree as much harm as good.

If birds or deer attack your crop the only remedy is to cover the tree with netting. You could leave windfalls for the animals instead.

TRAINING AND PRUNING APPLES AND PEARS

There are four common tree shapes: cordon, espalier, open-center bush, and dwarf pyramid.

Training a Cordon

This allows trees to be planted close together, making them easy to manage.

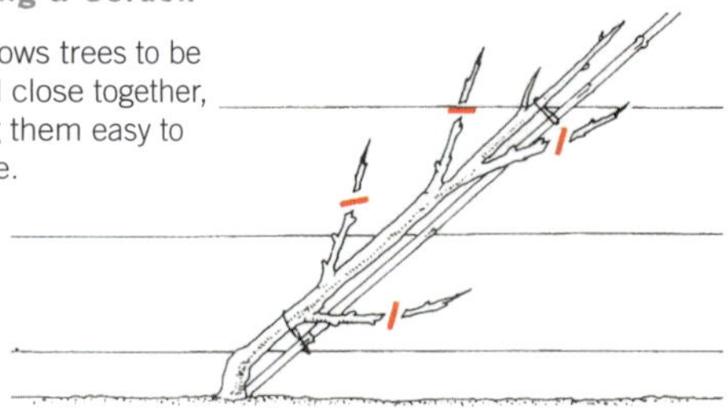

1 In winter, erect a posts-and-wire system. Plant the tree at an angle of 45°, the union uppermost. Tie trunk of tree to a bamboo stake, and tie that to the wires. Cut any sideshoots back to three or four buds. Do not prune the leader unless it is a tip-bearing cultivar, then prune this back by half.

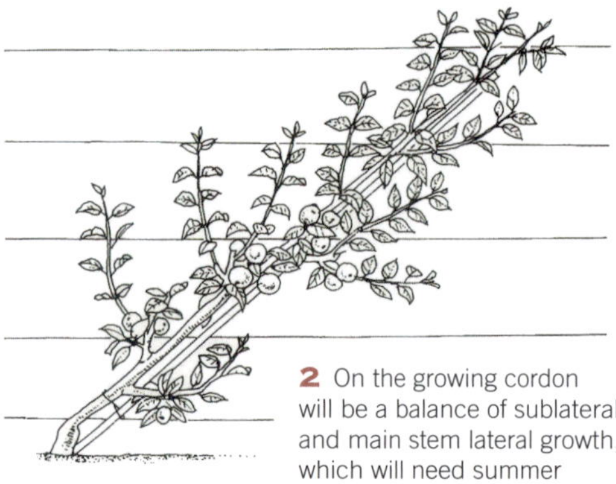

2 On the growing cordon will be a balance of sublateral and main stem lateral growth, which will need summer pruning, as below.

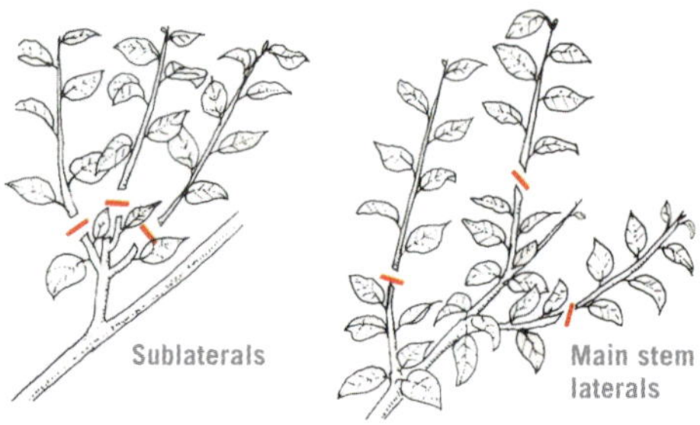

3 Start the summer pruning; cut back sublaterals to 1 in (2.5 cm) or one leaf, to encourage the formation of fruiting spurs. Prune laterals growing from main stem back to three leaves beyond the basal cluster. If the spur system becomes too crowded the spurs should be thinned in the winter by removing some and cutting back others.

Training an Espalier

An espalier is a decorative feature common in Europe, but it is also becoming a popular solution for small-space gardens in urban and suburban gardens in North America.

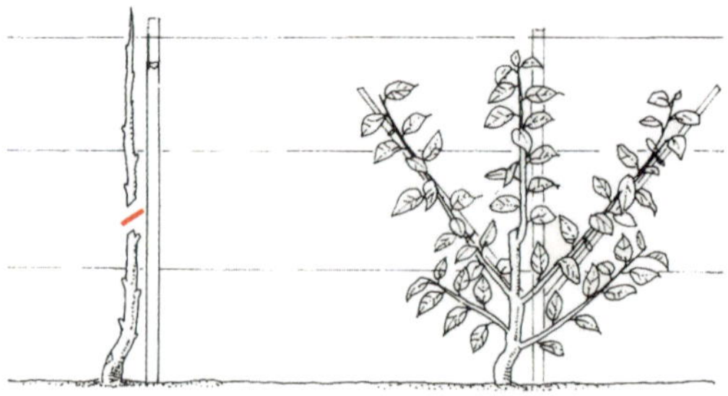

1 Erect stakes and wire. Plant tree in early winter, and cut back to a good bud with two buds facing outward beneath it, 12 in (30 cm) from the ground.

2 During summer, train the center shoot up the stake and tie sideshoots to canes set at 45° to the main stem. Secure the bamboo canes to the horizontal wires.

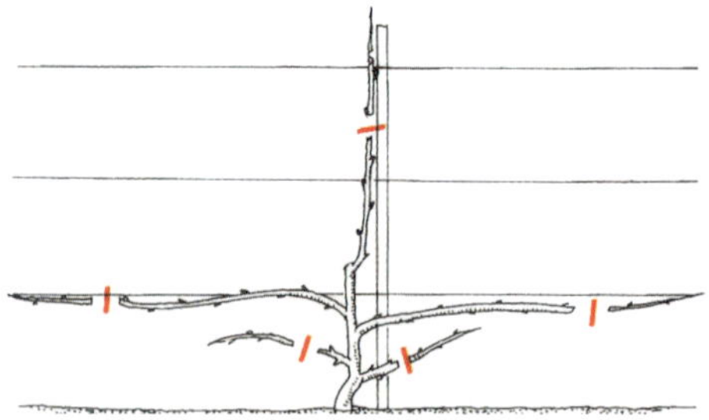

3 In winter, bend down side growths and tie to the horizontal wires. Prune them back to one third and the main stem to three good buds, two of which should face in opposite outward directions. These will form the next tier, 12 in (30 cm) from the lower one. Cut sideshoots down to three buds.

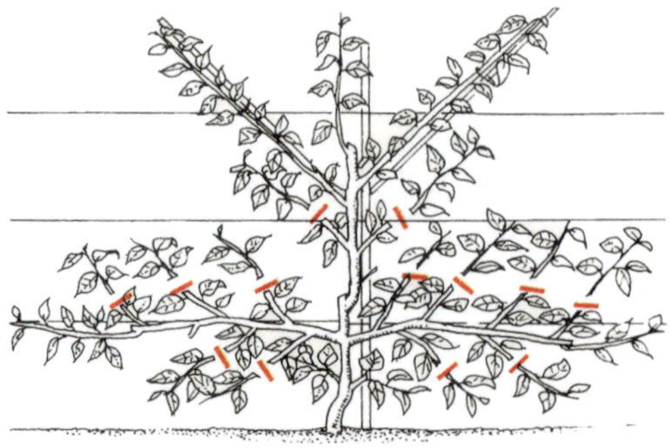

4 In summer, tie new tier to 45° canes. Start summer pruning on lower tiers by cutting back laterals on horizontals to three leaves to encourage fruiting spurs and cut sublaterals to 1 in (2.5 cm) or one leaf, like cordons. Cut growths from the main stem to three leaves. On established espaliers cut back main stem and horizontals to ripe wood.

Training an Open-Center Tree

This shape is used for growing apples, pears, peaches, plums, and cherries. However, it may take up to five years to train the tree into its required shape.

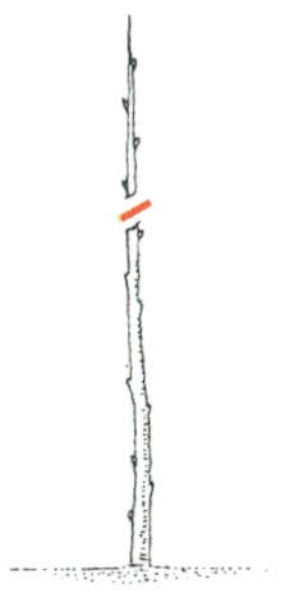

1 In winter, plant tree and cut it back to 2 ft (60 cm) at a point where there are four buds well positioned around it.

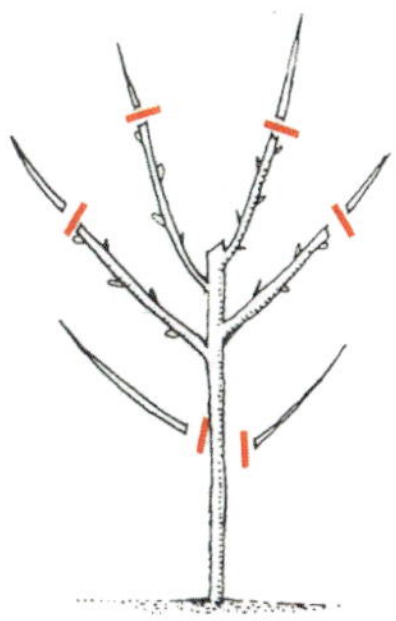

2 Next winter, cut back the four primary leaders by a third, to an outward-facing bud. Remove any other growths.

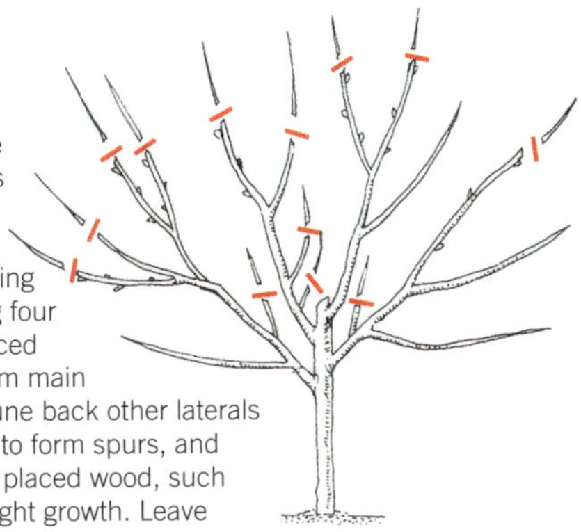

3 The following winter, cut the leading shoots by a third again, to an outward-pointing bud, selecting four more well-placed growths to form main branches. Prune back other laterals to three buds to form spurs, and remove badly placed wood, such as strong upright growth. Leave some outside laterals unpruned.

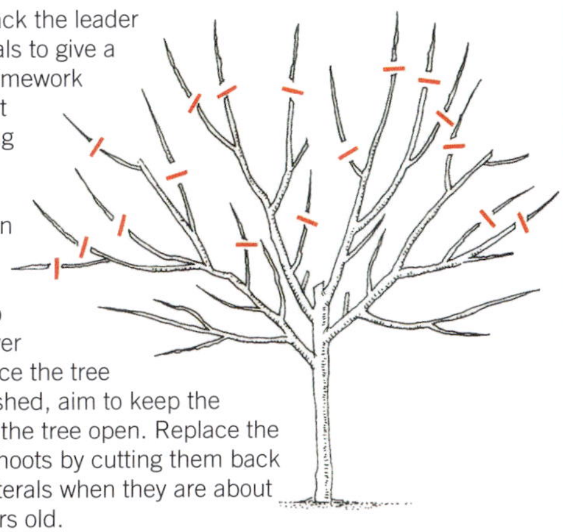

4 Cut back the leader and laterals to give a strong framework to support the fruiting growth and to strengthen it. Cut back the laterals to good flower buds. Once the tree is established, aim to keep the center of the tree open. Replace the leading shoots by cutting them back to new laterals when they are about three years old.

Training a Dwarf Pyramid

This intensive technique is suitable for dwarf and semidwarf trees. If planning to train fruit trees this way, you will need to stake young plants.

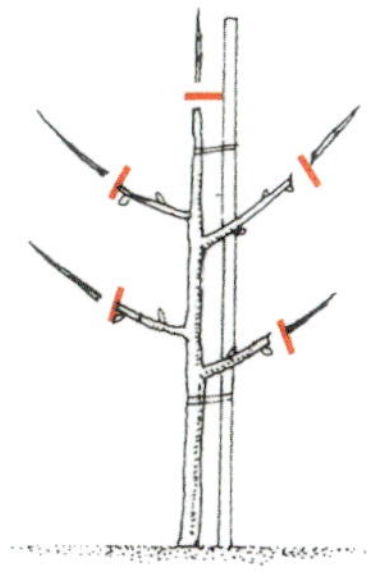

1 Plant in winter and cut back three or four laterals by one third to outward-facing buds. Laterals should be about 2 ft (60 cm) from the ground. Cut out other laterals, and cut back leader to two buds above the top lateral.

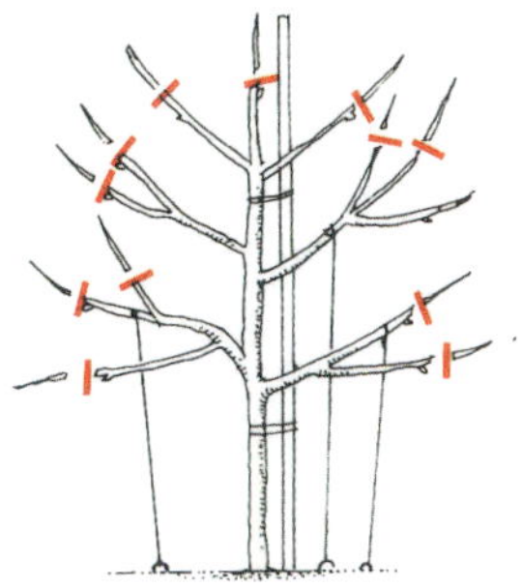

2 In summer, if growth of laterals has been good, tie them with soft string to 30° above the horizontal laterals.

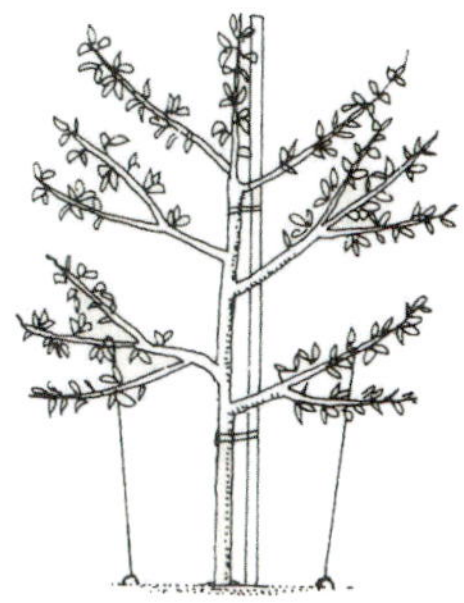

3 In winter, cut back main leader to an opposite growing bud. Remove any upright growth. Tip remaining laterals to a downward bud.

4 Remove ties when branch angle is set. Keep tying in new laterals and check that the string is not too tight and restricting. Incorporate new horizontal laterals but keep those at the top of the tree short to allow the sun to reach lower branches. Prune the central leader in winter. To maintain A-shape remove vigorous upright growth.

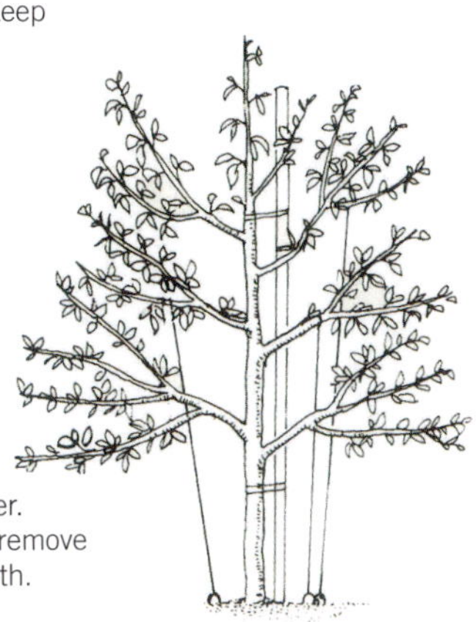

Growing Pears

The cultivation of pears is very similar to apples. Garden varieties are grafted onto dwarfing rootstocks of quinces, which produce larger trees than the smaller rootstocks of apples.

Pears grow best in a sunny, warm, and sheltered site. A south-facing wall is an ideal place for training a pear as an espalier.

Pears are not generally self-pollinating, although family trees of two or more varieties are available. You are likely to get better results, however, by growing two or more trees of different varieties. Varieties can be divided into those that mature early, midseason, or late.

Soil, position, and cultivation

Pears like a deeply cultivated loam that is moisture-retentive, particularly in the summer. They need a more sheltered spot than apples, and it should be warm and sunny. As they can bloom early, do not plant them in a hollow or pocket where frost collects. Pear trees do not grow well with grass around their roots, and the soil around them should be kept well-cultivated and perpetually free of weeds.

Early winter is the ideal time to plant. If you are planting more than one tree, leave approximately 13 ft (4 m) between them. Mulch the area around the roots in spring and make sure the trees are kept well watered in dry weather. They like more nitrogen than apples, and this should be supplied in spring.

Thinning out

Like apples, pears need some thinning; clusters should generally be reduced to one or two fruits as the developing fruits begin to turn downwards. Remove smaller fruits, leaving the healthiest in place.

Training and pruning

Pears are trained and pruned in the same way as apples (*see* pages 117–18), but, once established, can take harder pruning.

Harvesting

Early maturing varieties should be harvested while the pears are still hard (in late summer). They will not part easily from the tree, so cut stalks with pruners. If left on the tree, the pears will ripen unevenly. Lay them in trays and boxes and keep in a cool place until ripe.

Later varieties should be picked when they part easily from the tree and then ripened as above.

PESTS AND DISEASES

Pears are not vulnerable to many pests and diseases. Maintain good hygiene in your garden and prune the trees appropriately and you will avoid inviting problems.

The main pest is pear midge (page 149), although this is only a threat in certain areas. Pear leaf blister mites (page 145) can be a problem. Look for shriveled leaves or flowers that could indicate fireblight (page 146). Scale insects and aphids (page 144) may also be a problem.

SEASONAL GROWING GUIDE

Early to late winter	**Winter pruning** Cut back hard to invigorate new growth
Early summer	**Thinning** Pick off developing fruits to leave no more than two in each cluster
Midsummer	**Summer pruning** Cut out new growth to improve shape and avoid overcrowding
Late summer to late fall	**Harvesting** Pick when the fruit is still hard and ripen off the tree

Other Tree Fruit

Fruit trees take a longer time to become established than soft fruit and need more room to grow. However, they have a longer fruiting season and can also become a decorative feature in an urban or suburban garden.

Plums and cherries are hardy choices that will thrive in most areas and conditions, whereas peaches, nectarines, and apricots need more warmth and must be grown in a sheltered spot in northern zones.

PLUMS, DAMSONS, AND GAGES

These all belong to the same family, but vary in size and color, from small, deep-purple damsons to juicy Japanese plums that can be pinkish yellow as well as purple.

European plums typically grow best in cooler climates and Japanese plums are better suited to warm areas. Only a few are self-pollinating, so if you want to have only one tree, make sure you choose one of these.

Damsons are small, sour fruits, so they are generally cooked before eating. They make excellent jams and are delicious stewed or in pies. Gages may be eaten fresh, but are very often cooked; although they are sometimes called greengages, you can also grow yellow varieties.

Soil, position, and cultivation

All these trees will grow on most soils, but they like it to be well-drained. If it is wet and heavy, add lots of well-rotted organic matter when preparing the ground for planting. Avoid planting in a spot where frost collects—particularly for Japanese plums, which flower early. These trees will give their best crops if grown against a south-facing wall in cooler areas. European plums can withstand tougher conditions.

Planting essentials

Plant trees as early as possible in the growing season and if the soil is dry, soak it well the day before planting.

Keep the surrounding area free of weeds by hoeing, but try not to disturb the roots of the trees, as this will encourage the tree to put out suckers. If suckers do form, tear them off: cutting encourages further growth. Mulch regularly in fall.

Unless you want a glut of these fruits for jam-making or freezing, one tree of each type should be plenty for most gardens.

You don't need a lot of trees and certainly not a whole orchard to produce a wide variety of delicious tree fruit.

Growing a fan-shaped tree

For many tree fruits that originated in hotter climates, a south-facing masonry wall or fence can provide the warmth and shelter they need to fruit reliably. Most fruit trees can be trained to grow against a wall, where they will benefit from heat that radiates from the wall on cold nights following long and sunny days. Apricots, peaches, and nectarines are particularly well-suited to this style of training. Growing fruit trees in this way is also a very economical use of space if you have a small garden. Many other fruit and vegetables will not grow so well close to a high wall, so it is soil that would otherwise be left bare, and the trees themselves take up far less room like this than as trees grown in the open.

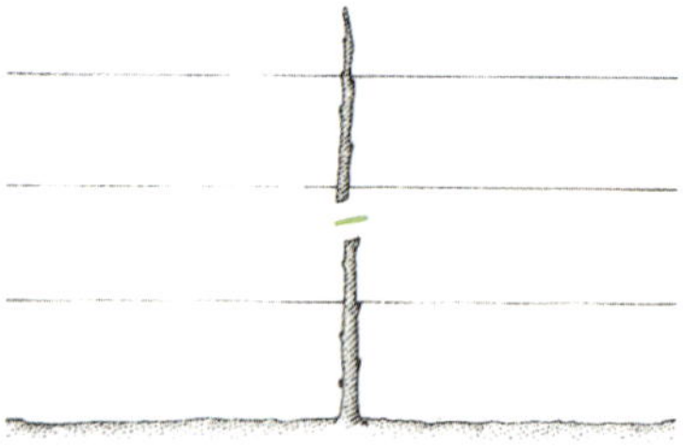

Training a fan In winter, erect a wire support system. Plant the tree and prune to three buds 12–18 in (30–45 cm) above ground.

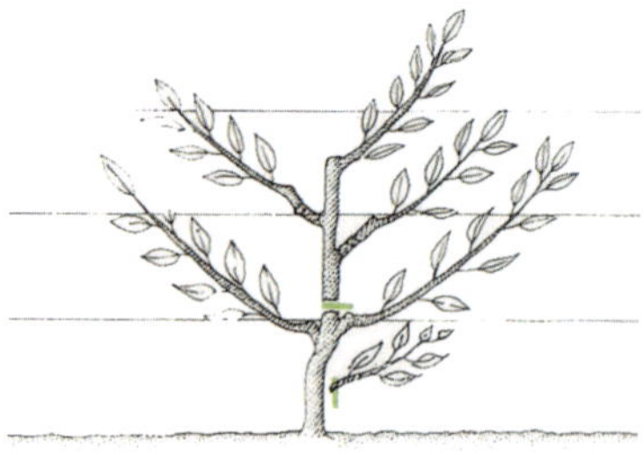

In summer Tie laterals to poles at 45° angles, which are attached to wires. Remove other sideshoots and cut and seal the main stem.

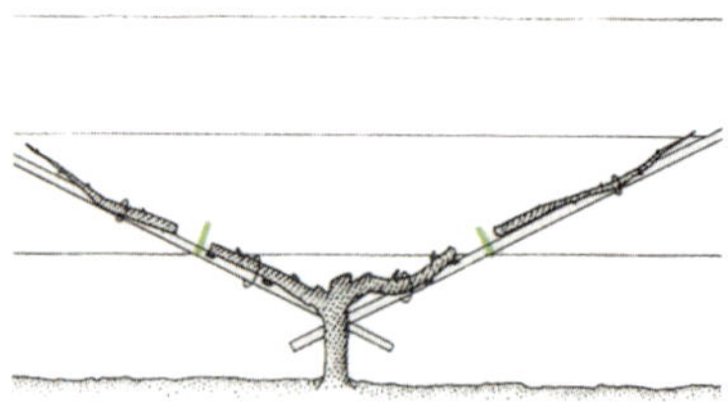

In winter Cut back the two laterals to give two good buds on the top, one at the end, and one underneath at the end.

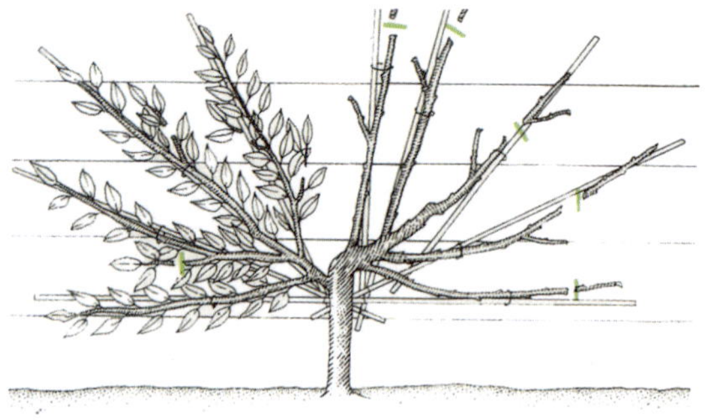

Next summer Train laterals to grow evenly spaced. Tie each shoot to an angled cane. Prune back sub-laterals to 3–4 in (7.5–10 cm).
In winter Cut back all leaders by one third to suitable buds.

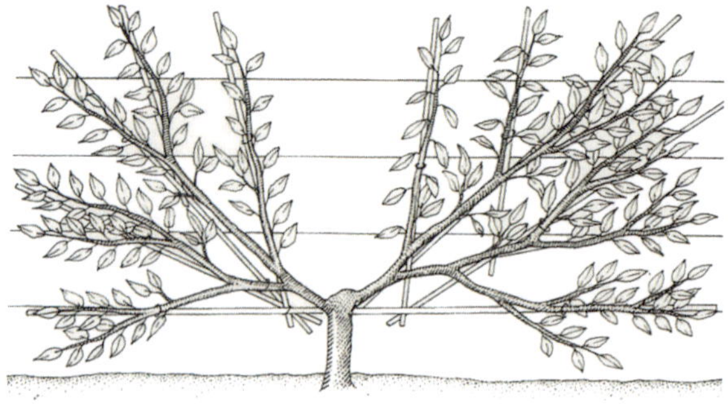

Established tree The tree is considered to have covered the wall space when it has a good framework of branches, each rib having a fruit-bearing lateral at 4 in (10 cm) intervals, every fourth one being a fruit-bearing lateral.

Pruning the fruiting fan Pinch back growth buds on fruiting laterals to two leaves, leaving a replacement and, if wanted, a reserve lateral.

Next summer Pinch back replacement laterals to 10 leaves, and pinch back fruiting ones to five, unless required for framework.

After harvesting Cut back fruited lateral to the best replacement lateral unless required in framework. Cut out old or unwanted shoots, always maintaining a balance of young and fruiting wood.

Training and pruning

Plums may be grown as open-center trees, pyramids, or as fans against a wall. As with other fruit trees, it is best to buy two- or three-year-old trees, in which the initial training has been done. The training for open-center bush and pyramid trees is the same as for apples (*see* page 117); train fan-shaped trees as described for apricots, opposite.

Once established, these fruits need very little pruning except to ensure that they do not become overcrowded. However, Japanese plums need to be more heavily pruned than European types. Once a year, remove very old, dead, or badly placed wood.

Do all pruning in spring to lessen the likelihood of attack by silverleaf (*see* page 149), which enters the tree through open cuts and wounds and which is at its least active at this time of year.

Thinning and harvesting

Plums, in particular, should be thinned, as too heavy a crop could cause the branches to snap, leaving open wounds that provide entry points for disease. Thin in two stages—first in late spring and about three weeks later. At this point, the fruits should be about 2 in (5 cm) apart.

Pick plums by the stalk to avoid bruising the fruit (the stalk should come away with the fruit as you pick). If you want plums for eating, leave them to ripen fully before picking, as they do not store well. For cooking, plums can all be picked before they are fully ripe. If there is a lot of rain at harvesting time, the skins of some plums are liable to split before they are fully ripe.

Pests and diseases

Plums are very attractive to hungry birds and the best protection against losing a lot of your crop is to cover the entire tree with netting or floating row covers if possible. This will also keep wasps away from the fruit.

Netting may not be pretty, but it is effective. The best way to fix and support the netting is to drive tall stakes into the ground around the tree—they must be a little taller than the tree itself—and drape the netting over and around them. Fix the netting to the stakes. Use a close mesh, as birds may get entangled in a wider mesh net.

Plums are prone to early frosts because they flower early in the year. Protect them, if possible.

Other common problems to watch out for are aphids (*see* page 144), plum curculio (page 148), and silverleaf (page 149).

Most fruit trees produce a wonderful display of blossoms in spring, at a time when the vegetable garden can look a little bleak.

PEACHES AND NECTARINES

These fruits originated around the Mediterranean, but may be grown successfully even in cooler regions.

The nectarine is a type of peach, with smooth, rather than velvety, skin. It is also less hardy. Both are self-pollinating, although it will help if you brush over the flowers with a soft paintbrush around midday on sunny days.

Soil, position, and cultivation

Both trees like well-cultivated, well-drained soil, which will retain moisture in summer. They need some winter chill, but they cannot survive very cold winters, so consult your local Cooperative Extension Office for advice on varieties in your area. Growing fan-trained trees against a south-facing wall or fence is a good option in cooler climates.

Plant as early as possible in the recommended planting period. Keep the ground around the roots free of weeds, but do not hoe too deeply in case you accidentally damage the roots. Water well in dry weather and apply a mulch of well-rotted compost or manure in spring.

Pruning and training

In cooler zones, train as fans on a wall (*see* page 120). Once established, prune to get a constant supply of one-year-old wood, as peaches and nectarines produce fruit on the previous year's wood.

Thinning and harvesting

The crops of both fruits must be thinned to produce decent-sized fruits. Do this in two stages—when the fruits are marble-sized, reducing clusters to single fruits 4 in (10 cm) apart, and again when they are a little over walnut-sized. At this stage, reduce the crop by half so that they are about 9 in (25 cm) apart. Nectarines should be thinned to 6 in (15 cm). In both cases, at the first thinning remove all fruits that are growing towards the wall and will not have room to develop properly.

Test for ripeness by supporting the fruit in your hand and very gently pressing the flesh by the stalk. It should give and the fruit will come away easily from the tree. Peaches and nectarines bruise easily.

Nectarines (above) and peaches are grown in the same way; they are self-pollinating, so you will only need one tree.

SEASONAL GROWING GUIDE

FRUIT	SOIL AND POSITION	PRUNING	THINNING FRUIT
Plums	Most well-drained soils. Avoid frost pockets	Very little required. Remove old, dead, or badly-placed wood in spring or summer	Late spring and early summer to 2 in (5 cm) apart
Peaches and nectarines	Well-cultivated, well-drained soil. Need warm conditions	Prune to maintain a fan shape and a good supply of one-year-old wood	Thin in two stages to 9 in (25 cm) apart
Cherries	Deeply-dug and well-drained. Any position	Prune to maintain shape and encourage a constant supply of new growth. Cut out old wood	No thinning required
Apricots	Well-drained, fertile loam in a warm and sheltered position	Prune into a fan shape, trained against a wall	Thin to 4–5 in (10–12.5 cm) apart

CHERRIES

There are two types of cherry: sweet and sour. Sweet cherries are rarely suitable for small gardens as they are not normally grown on dwarfing rootstocks and the full-sized trees are very large. Duke cherries are like a cross between sweet and sour cherries.

It is possible to fan-train sweet cherries against a south- or west-facing wall, but they will grow up to 20 ft (6 m) or more in all directions and, as very few varieties are self-pollinating, you will need more than one tree. Also, unless they are very closely netted, the birds will take the entire crop before you get a chance.

Sour cherries, of which "Morello" is the most popular, are a better choice in small gardens, and may be grown as bushes, trees, or against a wall. They are self-pollinating and are not subject to the same dedicated attack from hungry birds as sweet cherries are.

Soil, position, and cultivation

Because cherries send their roots way down into the soil, they like ground that has been deeply dug and is well-drained. Add lots of well-rotted organic matter when preparing the ground. If there is any danger of drying out, mulch around the roots well in early spring.

Pruning and training

Sour cherries can be trained as described on pages 117 and 120. In subsequent years, prune to get a constant supply of new growth, as sour cherries fruit only on the previous year's wood. Sweet and duke cherries fruit on older wood—two years old and older. Prune cherries in spring to avoid disease.

CITRUS

Gardeners in zones 10 and 11 think nothing of having an orange or lemon tree in the garden. Elsewhere, citrus trees will need winter protection and even then may not survive a prolonged period of harsh weather.

Even if you don't live in California or Florida, you can still grow citrus trees. If you have a greenhouse, a sunny, sheltered patio, or a sunroom, you can keep them out of the cold. Never allow the potting soil to dry out or the plant will progressively lose its flowers, fruit, and then the leaves. Water regularly all year round and mulch around the base of the trees with light-colored gravel.

The fruits take 12 months to mature, even longer for grapefruits, and for them to ripen, the trees need at least six months after flowering when the temperature stays continuously above 61°F (16°C), so even indoors the trees may need some extra heat. The heat is more important than actual sunshine for citrus to fruit.

Container-grown oranges, lemons, limes, and grapefruits can all be bought at good garden centers, usually all year round. You might also find some more unusual crossed varieties, such as the orange and lime cross, bergamot.

Citrus trees are grafted onto rootstocks, so be sure to remove any suckers that emerge from below the graft union.

Harvesting

There is no need to thin cherries. Harvest when they are ripe by cutting off the stalks, using scissors. If you harvest sour cherries by pulling them off the trees (as you can with sweet cherries), disease may enter where the bark is torn.

APRICOTS

Apricots can be grown outside in warm areas, but in very cold areas they will really only survive and produce decent yields of fruit if grown indoors. They can be trained against a wall or fence in colder regions (*see* page 120).

Soil, position, and cultivation

Apricots like well-drained, fertile soil, preferably a fairly light loam. A warm, sunny position is essential, ideally against a sheltered, south-facing wall. Apricots may be planted in early winter or midwinter. If planting early, wait until the following spring to begin pruning.

Apricots are self-pollinating, but as they can flower in late winter before insects are active, it is a good idea to pollinate them by hand. Protect buds from frost by covering the tree with floating row covers.

PESTS AND DISEASES

Peaches, nectarines, cherries, and apricots are all susceptible to the same few pests and diseases.

Inspect trees regularly for signs of infection or pests and treat any problems promptly. In particular, look out for bacterial canker (page 144), peach leaf curl (page 145), stink and tarnished bugs (page 148), and brown rot (page 148).

Growing Figs and Grapes

Both these fruits originated in the sunny Mediterranean and love long, hot summers.

Grapevines will take a few years to get established but, in time, can produce enough grapes to make it worth trying your hand at making your own wine (*see* pages 230–33).

FIGS

In colder districts, figs may need to be grown in a greenhouse, but in many areas there is no reason why a fig tree should not thrive, especially if you have a sheltered patio or courtyard garden.

Figs grow best in poor soils, ideally against a south-facing wall, but if they are sheltered, they can fruit prolifically. Unlike other tree fruits, figs are not grafted onto a different rootstock, so their roots must be confined to produce a good crop of fruit. Do this by lining the planting hole with stone, or concrete slabs, bricks, or aluminum sheet, or by planting a tree in a large barrel with holes pierced in the bottom for drainage. Soil mixed with gravel and bonemeal is a good planting mix.

Plant trees in spring. If planting more than one tree, allow 15–18 ft (4.5–5.5 m) between them, although one tree is usually sufficient.

No mulching or amending will be needed in the first few years as this will encourage over-vigorous growth. Once established, a mulch of well-rotted compost or manure will be beneficial in dry conditions.

Training and pruning

A fan-shaped tree may be trained as shown on page 120 and a freestanding tree as on page 117. Once established, cut back sideshoots in early summer to 4 in (10 cm), or so that about five leaves remain. Remove suckers and unwanted growth after one month, and tie in new shoots to keep the overall shape of the plant.

Figs are ripe and ready to harvest when their skins have turned brown and started to split and the fruit feels soft to touch.

Harvesting

Figs may produce a harvest up to three times a year, depending on the climate. In cooler areas, they will form two crops but usually only one will ripen fully. In warm regions, they may crop up to three times a year.

In cooler climates, discard the small, early-summer fruits in fall. Over the winter, the late summer fruits develop and ripen the following summer and fall. Harvest the larger fruits as they ripen—the stem will soften, the fruits turn brown, and the skins start to split.

GRAPES

Grapes can be grown outdoors or in a greenhouse (*see* page 127). Both table and wine varieties—either white or red—are available.

You will only produce a really good crop if your vines have lots of warm summer sun. Choose early ripening grape varieties for colder districts. It's important to choose the right variety for your region, so check with your local Cooperative Extension Office for advice. Ideally grapevines like sandy, stony, or gravelly soil that is neutral or slightly alkaline. It must be well-drained.

Grapes need a sunny, sheltered position where there is no danger of frost lingering. A south-facing wall or fence is ideal. Prepare the site by digging in a little well-rotted compost. Buy one-year-old vines and plant them in spring, about 5 ft (1.5 m) apart. Keep the ground moist. The amount you grow is likely to be dictated by available space. Even in a small garden, an established vine grown against a wall can produce up to 20 lbs (9 kg) of fruit, or more.

Training and pruning

The easiest way of training vines is to grow the vine over an arbor or patio roof, or to grow a single cordon on a wall. Vines should not bear fruit until they are three years old; they bear grapes on the current year's growth.

After planting, cut out all but the strongest shoot and tie this to a stake. Pinch out the flowers as they appear in summer, and cut back laterals to five leaves. Tie these laterals to wires. Rub out laterals that grow between the wires.

Repeat this process the following year. The next year, allow the vine to produce three bunches of grapes, and in the next year four or five bunches. Once established, prune the vine after the leaves have fallen, as below. Remove rotting berries as you spot them.

PESTS AND DISEASES

Figs and grapes suffer from few problems, but birds, wasps, and diseases are most common.

Be alert for signs of armillaria root rot (*see* page 148) and coral spot (*see* page 147) before they spread too far. Armillaria can kill a vine; dig up and discard the plant and change or sterilize the soil before planting a replacement in the same place.

PRUNING AND TRAINING A GRAPEVINE

Training a vine In winter, erect wire supports and plant vine. Cut main stem by two thirds. Cut laterals back to one bud.

In summer Remove flowers. Cut laterals back to five leaves and tie loosely to wires. Pinch out sublaterals to one leaf.

Next winter After leaves have fallen, cut vine back by two thirds and prune laterals on the main stem back to a good bud. Repeat process next year.

Third summer Train leader to grow along top wire. Only allow two laterals to fruit and pinch out to two leaves beyond fruit.

Greenhouse Fruit

A greenhouse can help you to get better and more reliable crops from tender fruits, such as melons, grapes, peaches, and nectarines.

Greenhouses are the only way for gardeners in coldest zones to grow heat-loving crops. Even a small garden can find space for a greenhouse to grow exotic fruits. If you are aiming for variety it is worth including as much greenhouse space as you can.

MELONS

Melons can be grown in a cool greenhouse, but ensure they are producing fruit and not just foliage.

As long as you keep melons well-watered and pinch out the sideshoots, there is no reason why you should not be able to grow your own fruit. Look for small cantaloupe and honeydew varieties recommended for growing indoors.

Sowing and planting out

Raise the seed in a propagation unit, sowing in spring or earlier if your greenhouse is heated. Prick out into small pots, then plant seedlings in their permanent position in the border soil when they have four or five leaves. Plant 2 ft (60 cm) apart and construct a supporting framework similar to that for cucumbers (*see* page 102). Pinch out the growing tips—the main stem should have no more than about six large leaves—and train the sideshoots along the horizontal wires. Pinch out their growing tips when they are 1ft (30 cm) long.

Melons will need pollinating by hand; transfer pollen from the male flowers (those with no embryonic fruit behind the petals) to the female flowers using a fine paint brush or by pressing the male stamen into the female flowers.

Routine care

Keep the plants well-watered and feed regularly with liquid fertilizer as the fruits begin to develop. If the plant seems very heavily loaded with fruit, pick out any that are either very small or very large so that the fruit

In coldest zones, figs and grapes will crop more reliably and in larger quantities when they are grown in a greenhouse.

left behind is a uniform size. Stop watering and feeding the plant when the melons have stopped swelling.

Support the melons during the ripening phase by placing them on a board or inside some netting suspended at an appropriate height. When they are ripe they will feel just soft when you push them gently by the stalk and will smell delicious.

Watermelons

You can grow watermelons in a heated greenhouse in the same way as for other melons. The fruits are around 95 percent water, so regular, plentiful, and even watering is critical for success—only stop watering while the fruit ripens.

The fruits are exceptionally heavy, but the plant grows by sprawling across the ground, not climbing up supports. Raise the ripening fruits on wooden boards, low stools, or in a net suspended from a sturdy A-frame to keep them off wet ground that could cause them to rot. Water watermelons at the base, not from above, which can encourage mildew.

GRAPES

Vines grown in even and unheated greenhouse are likely to yield more grapes than those grown outdoors, especially if they are planted in a south-facing position.

If you have a lean-to greenhouse on a south-facing wall, you could grow vines on the greenhouse wall furthest from the house, reserving the house wall itself for tree fruits. Keep the vines' growth in check, so that they do not exclude too much light.

Planting and training

Vines can be planted either in the greenhouse in-ground beds or just outside, trained in through a small hole in the wall. The simplest way to grow grapes in a greenhouse is on a single-stem cordon or rod—the same system as described for outdoor cultivation (*see* page 125).

Vines are usually bought as one-year-old plants, and planted in fall. Plant in fertile, damp soil, enriched with well-rotted compost. To train a single stem, tie it to a stake and pinch out laterals when they have five to six leaves. Train these along wires or strings.

Rub out any sideshoots that grow from leaf axils and stop the lead shoot when it has reached the height you want or is level with the bottom of the greenhouse roof. Do not allow the plant to bear fruit until it is three years old. Even then, restrict cropping to two bunches for each lateral. Vines should be pruned hard each winter, taking the laterals back to a new bud close to the main stem.

Thinning out and feeding

As the grapes develop, cut out the small grapes from each bunch. This allows for bigger ones to develop and prevents the smaller ones turning moldy as they get crushed by the larger ones.

Remember to water the vine occasionally during winter and topdress the soil with fertilizer and well-rotted compost in spring.

TREE FRUITS

Peaches, nectarines, apricots, and figs will all thrive indoors and against a south-facing wall.

Although these fruits can all be grown outdoors (*see* pages 122–25) they will also crop in a greenhouse. The general care and cultivation for growing tree fruits in a greenhouse is just the same as for growing outside, but they must be planted in very well-drained soil, which is fertile even at quite a depth—peaches extend their roots down for about 2 ft (60 cm). The trees must be kept well-watered throughout the year, even in winter, when there appears to be little or no growth. Give them a good mulch of well-rotted compost or manure each spring.

Balance warmth and fresh air

In winter, make sure that you shut all ventilation windows early in the afternoon to trap any warmth in the greenhouse. Open them in the morning to let in some fresh air. As the weather warms up, shade the greenhouse with blinds or shading paint to prevent scorching in sun.

Pollination is less likely to occur naturally in a greenhouse, because insects do not have free access to the blossoms, so you will need to do this by hand. Brush over the flowers with a soft paint brush, transferring the pollen from flower to flower. This is best done around midday.

Routine care

Indoor tree fruit needs spraying or syringing with water to ward off red spider mite and to encourage the flowers to set fruit. Spray once a day when the flowers are out and twice a day when the fruit buds have begun to emerge. Never spray the plants in direct sun and stop when the fruit begins to ripen.

Training and pruning are the same as for the same plants grown outside. Train trees as fans against a wall to be as efficient as possible with space and prune hard in spring so that the developing fruit is not obscured by foliage. If there is a lot of foliage, tie it back to allow sunlight to get to the ripening fruit.

Growing Rhubarb

Rhubarb is ready for picking from early spring onwards, earlier if it is forced, at a time when virtually no other fresh fruit is available in the garden. It can be used in a variety of desserts.

Only the stems of rhubarb are edible. Harvest flower stems as they emerge and put the leaves in the compost pile.

Rhubarb will grow in just about all garden soils and sites, but does best in an open, sunny position. It suffers from very few pests or diseases.

Planting and routine care

Because rhubarb is a perennial and can stay in the same place for five years or more, it is worth preparing the bed well. Dig it over and remove all perennial weeds, then incorporate some well-rotted organic matter, digging it deep, because rhubarb sends down long roots.

Plant rhubarb crowns in early spring so there is about 3 ft (1 m) between plants. The holes should be wide enough to take the entire rootstock and deep enough to allow only small shoots to protrude from the surface. Firm the soil around the rootstock and water if it is at all dry.

Water well in dry spells and mulch the plants each spring. Feed them in summer with liquid fertilizer. Cut out any flower stems and do not pick any rhubarb in the first year.

How many plants to grow

An established rhubarb plant should yield about 10 lbs (4.5 kg) of fruit in a year, but remember this is spread over a period of nearly six months.

Harvesting

Pull stalks from the plant when they are turning pink. Hold them near the bottom and pull them from the plant with a gentle twisting movement. You should be able to gather rhubarb from late winter through summer, but do not pick too heavily in the second year and always leave a few good stems and leaves on the plant to help it gather its resources.

Raising new plants

New rhubarb plants can be obtained by digging up the old plant and dividing up the roots into pieces that each have a growing bud. Do this in early spring using two- or three-year-old plants.

SEASONAL GROWING GUIDE

Planting	**Early spring** After preparing the ground in winter
Forcing	**Late winter** Outdoors **Early winter** Indoors
Harvesting	**From early spring** Plant out rhubarb forced indoors for another crop

Forcing rhubarb

Rhubarb is generally ready for harvesting in spring, but it can be forced to be ready for picking earlier. To do this, cover the emerging leaves with a bucket or large container, making sure no light can enter. You can also use traditional clay or terracotta rhubarb forcers (below), which are an elegant shape—but they are expensive and hard to find. A bucket will do the trick just fine. You can also dig up the plants and bring them into a cool, dry place in early winter, so they are ready for harvesting in midwinter. Then plant them outdoors again and cover with a bucket or pot to force. The following year, leave the plant unpicked so it can build up its reserves again.

Forced rhubarb stems are pale and tender. Plants grow up fast to reach the light at the top of the forcing pot.

Growing Strawberries

Strawberries differ from other soft fruit in that they grow as small perennial plants, rather than on bushes or canes. Although considered to be one of the greatest fruit treats of all, they are not difficult to grow and it is possible to produce them in most gardens from late spring right up to the first frosts.

SEASONAL GROWING GUIDE

VARIETY	PLANTING	HARVESTING MONTHS
June-bearing	**Late summer–mid-fall** If you cannot plant until spring, pinch off flowers and delay fruiting for a year while the plants establish	Protect fruit with straw from late spring. Harvest late spring to late summer, depending on variety
Everbearing	**Late summer** Allow 18 in (45 cm) between plants	Protect fruit with straw from late spring. Harvest from early summer into fall
Alpine	**Fall–midspring** Sow seeds over winter. Transplant young plants later in spring, 12 in (30 cm) apart	Harvest summer to fall

There are three types of strawberries: June-bearing types that produce one heavy crop—either in spring or summer; everbearing types, which produce a number of crops from summer to fall; and alpine strawberries, which bear their tiny, highly succulent fruits during the summer. They are all good sources of vitamin C.

Although strawberry plants are perennials, they should be pulled up and replaced after two or three years. After that, they produce increasingly poor crops and are also more susceptible to disease.

You can buy your first June-bearing and everbearing strawberries as bare-root plants, and then propagate the runners to get replacement plants. Alpines are best treated as annuals and raised from seed sown in the spring or grown from young plants.

Soil and position

Strawberries grow in all well-cultivated garden soils, but they do best in those that are well-drained and very rich in humus. Dig in lots of well-rotted organic matter about a month before planting. An open, sunny position will produce the heaviest crops, although they will grow successfully in some shade.

If space is a problem, strawberries can be grown in tubs, boxes, barrels, special strawberry pots, or among the flowers in a border. Strawberry pots have holes in the bottom for good drainage and a layer of large pebbles should be put in the base before adding the potting soil. Water regularly to keep the soil moist.

Planting and routine care

If buying new strawberry plants, choose only those that are certified disease-free. They can be planted any time from midsummer to early fall, to fruit the following summer.

If planting has to be delayed until the following spring, pinch off the flowers as they appear to allow the plant to direct its resources into becoming established before fruiting the following year.

Strawberries grow well in pots on a sunny patio, as long as they are well-watered. Raising pots off the ground can help to protect them from some hungry garden pests.

RAISING NEW STRAWBERRY PLANTS

1 Select one or two healthy runners from each plant in summer, and pinch out all the other runners where they join the parent plant. They will take up precious energy.

2 Sink a small pot of potting soil for each runner into the ground close to the parent plant and pin the runner into it. Pinch out any more growth to encourage the runner to root, rather than to grow.

3 Cut the runner from the parent plant when it is well-rooted and plant it out in its new site.

CULTIVATION OF STRAWBERRIES

1 Make a hole in the soil deep and wide enough for the rootball. Make a low mound of soil at the bottom of the hole. Position the plant in the hole, spreading out the roots over the mound of soil.

2 Protect growing plants from wet soil, which will cause them to rot, by growing them on special strawberry mats (above) or over a mulch of straw.

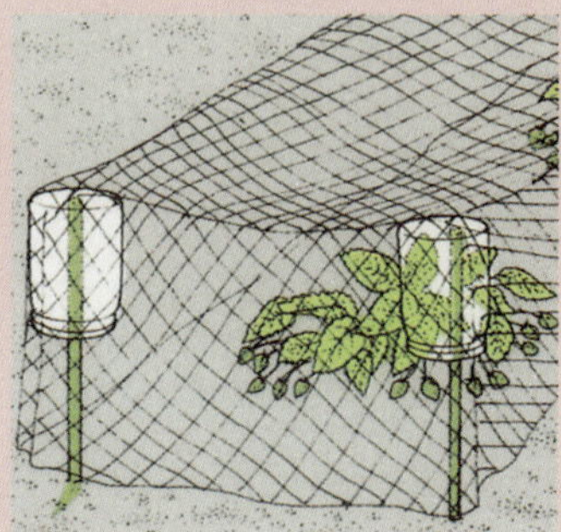

3 When fruits begin to form on the plants, cover them with cages of netting to deter birds from eating the strawberries.

Water plants well immediately after planting and for the next few weeks. Always make sure they do not dry out. Keep weeds under control by shallow hoeing, but take care not to disturb the plants' roots.

If you are growing strawberries in a greenhouse, you will need to pollinate the flowers by hand, daily, as soon as they open. Use a small, soft paint brush to transfer pollen from flower to flower.

Taking care of developing fruit

When the fruit begins to appear, put a 2 in (5 cm) layer of clean, weed-free straw between rows and tuck it well under and around each plant. This helps to protect the fruit from soil splashing up in wet weather and rotting it. Alternative methods of protection are to use special strawberry mats sold at garden centers, to lay black or red plastic mulch on the ground, or to slide empty glass jars over the fruits and rest them on the ground. Make sure the soil is moist if using plastic.

Protect developing fruit from attack by birds by covering the plants with netting supported on low stakes driven into the ground. Give applications of liquid fertilizer to increase the size of the fruits once they begin to swell. Pinch off runners as they appear unless you want to raise new plants.

If you want early crops, protect the plant with hoop houses or glass cloches in late winter. If you want late crops from everbearing varieties, cover in early fall. You should get crops at least until late fall.

At the end of the season

After harvesting, strawberry plants may be treated in a number of ways. With single-crop plants, loosen the straw around the plants, trim

Protecting strawberries

One of the best methods of protecting growing fruits is to raise them on beds of straw. Not only will it prevent the roots from rotting on soggy ground in wet weather, but when the season is over, the straw can be pulled up, taking any strawberry pests with it.

off dead leaves, and compost the debris. This removes old leaves as well as destroying pests and any mold spores that have collected in the straw. An alternative, and the best solution for everbearing varieties, is to cut off all the foliage to about 4 in (10 cm) above the crown of the plant, leaving the young growth to come through. Also remove the old straw.

A third way is to take one runner from each plant, and root them (*see* opposite) to form a new row. Then dig the old crop into the soil, or dig up the plants and dispose of them. This way you can produce a new row of strawberry plants every one or two years very easily.

To raise alpine strawberries, sow the seeds in potting soil indoors in late summer and prick out in early fall, or raise indoors in midspring. Plant out the seedlings in late spring, setting them 12 in (30 cm) apart, with the same distance between each row. After that, treat them as for other strawberries.

Harvesting

Pick strawberries on a dry day, when they are fully ripe, pinching them off by the stalk if you can to avoid handling and bruising the delicate fruit. Pick alpine strawberries throughout the season to encourage the plants to continue fruiting.

To lift fruits off the wet soil—keeping them clean and minimizing the chance of rotting—put down a layer of straw beneath the plants.

Pests and diseases

Aphids can be a problem on strawberry plants (*see* page 144). Slugs and snails are partial to young fruits; you can put down slug pellets before laying down a mulch of straw or plastic to protect the fruit. Handpick slugs and snails at night, too, for best control. Look out for signs of virus diseases (*see* page 146), weevils (page 148), mildew (page 145) and botrytis, or gray mold (page 146).

Good strawberry varieties

JUNE-BEARING
- **Benton** Heavy cropper, scarlet; good for Northwest
- **Sequoia** Developed for coastal areas; long crop

EVERBEARING
- **Fort Laramie** Very cold hardy and adaptable; excellent flavor
- **Tristar** Large fruits, mildew resistant, widely adapted

ALPINE
- **Mignonette** Masses of medium-sized bright fruits.

Other Berries

Raspberries grow in almost all zones, except the hottest, but they do prefer cooler climates. Currants and gooseberries are less widely grown than other berries; they also do best in areas with cool growing seasons.

SEASONAL GROWING GUIDE

VARIETIES	PRUNING	HARVEST
Summer-bearing Cumberland, Heritage, Tulameen	Cut fruiting canes down to ground level after harvesting. Leave the strongest new growth to bear fruit next year, cutting each cane back to a bud in early spring	After fruiting in summer
Everbearing Fall Gold, Summit	Cut fruiting canes back by one-third in fall; then again after fruiting in summer	Fall, and again in early summer

RASPBERRIES

Raspberries are very easy to grow and are among the most popular of all soft fruits. There are many different varieties—even some with yellow fruits—but they can be divided into two distinct types. Summer-bearing raspberries usually crop in summer, bearing their fruit on the canes produced in the previous year; everbearing types give two crops in a single year; the first in late summer and fall, then a second crop the following summer.

Soil and position

Raspberries will grow in almost all well-cultivated soils, but they prefer those that are slightly acidic and well-drained, but moisture-retentive. They will grow in partial shade, but do best in a sunny site that is sheltered from strong winds.

Planting and care

Dig lots of well-rotted manure or compost into the ground a month or two before planting, and make sure that the site is completely free of perennial weeds. The canes can stay in the same site for years, so take care over the preparation of the ground. Plant, preferably in fall, as described on page 112, leaving 18 in (45 cm) between plants and 5–6 ft (1.5–1.8 m) between rows. Buy only certified, disease-free stock—anything else may be harboring virus diseases.

Water the canes well during summer and apply an annual spring mulch. Keep weeds under control by shallow hoeing, but be careful not to disturb the roots, which run close to the surface—a thick mulch may be a more effective solution to eliminate weeds. Protect the plants and the fruit from birds by covering them with netting or by growing them in a fruit cage.

Established canes bear approximately 1 lb (450 g) of fruit each year. Do not allow canes to fruit in their first season: cut out any flowers that appear to prevent them from producing berries.

For support, tie the canes to horizontal wires strung between sturdy posts hammered in at either end of the row or to single posts positioned at intervals along the row (*see panel,* right).

After harvesting in the second season, cut out the fruiting canes on summer-bearing varieties just above the soil. Tie the eight strongest canes to the wires, and cut out all the weaker ones. Remove unwanted suckers that have sprung up between the rows at the same time. In very early spring, cut canes back to a bud that is no more than 6 in (15 cm) above the wire. Cut some canes a little shorter so that you have fruit at all levels. Cut everbearing varieties back by a third after fall fruiting, then down to the ground after the next summer crop.

SUPPORTING RASPBERRY CANES

Raspberry canes cannot grow without support. The method you choose may depend on the size of your plot and the position of your canes; try one of these three tried-and-tested solutions.

- Drive stakes into the ground at either end of the row of canes and string parallel lengths of wire between the posts. Put the lowest wire around 12 in (30 cm) from the ground. As the canes grow, tie them to the wires.

- Drive stakes into the middle of each group of canes. Loosely tie the canes to the stakes as they grow. Use this method to support raspberry canes grown in pots.

- Make a box structure to support the canes by driving one stake into either end of each row. Attach two cross struts to each stake and string wire between these, enclosing the rows of canes within a wire "box."

Horizontal wires strung between stakes are a simple way to support raspberry canes. Cut back new growth to just above the level of a supporting wire.

Prune summer-bearing canes that have borne fruit, cutting them back to ground level; select the strongest new growth and train it on wires for next year's harvest.

Harvesting

Pull the fruits off the canes gently as soon as they are ripe, leaving the stalk and core behind. They should be used immediately.

Raising new plants

New canes are easily produced from the suckers put out by existing plants, but only take these from certified virus-free canes. Dig them up and transplant in fall. Buy new plants every eight years or so.

Pests and diseases

Maggots of the raspberry beetle (*see* page 149) tunnel into the fruits, making them distorted and inedible. Look for and pick off the beetles, which lay their eggs on the flowers, and the brown grubs that follow. Raspberries are also susceptible to various fungal diseases, including cane spot (*see* page 147), which causes purple spots to appear on the canes. These later turn to white and the cane eventually splits. Spur blight (*see* page 147) also causes purple spots on the canes, followed by withered leaves; in bad cases, the canes may snap. Cut out and destroy infected canes to prevent the problem from spreading.

BLUEBERRIES

Blueberries are widely grown, but you must choose the right type for your garden. Highbush blueberries are widely adapted; lowbush types are under 3 ft (1 m) high and grow best in zones 3 to 6. In the hottest parts of the South, grow rabbiteye blueberries.

Soil and position

Blueberries like an acidic soil and will not grow on those which are alkaline. In these areas, you can still grow blueberries in large containers filled with an acidic potting soil.

Peaty, moisture-retentive soil will produce the best crops. Although lowbush and highbush blueberries can withstand cold conditions, they need sheltering from strong, cold winds. An open, sunny site gives the best crops.

Planting and routine care

Dig some acidic planting mix or peat moss into the site a month before planting, and plant as described on page 111. Blueberry bushes are best started as three-year-old plants bought from a local nursery, and plant them at the distances recommended for the type and variety. They will appreciate an annual mulch such as pine needles or bark mulch in early summer.

Blueberries will produce anything up to 10 lbs (4.5 kg) of fruit per bush, depending on weather and growing conditions.

Blueberries need no pruning until they are four years old. After that, cut out old wood regularly each winter, taking it back to the ground or a strong new shoot. Cut out any suckers that emerge at the base of the plant.

SEASONAL GROWING GUIDE FOR BLUEBERRIES

Planting	**Late fall–early spring** Space plants as recommended
Pruning	**Late fall–early spring** Cut back old stems to the ground or a vigorous new shoot
Harvesting	**Summer–fall** Pick over the bush several times, selecting the berries that are ripe

Pests and diseases

Blueberries are generally free from problems and pests or diseases.

BLACK CURRANTS

Black currants are best grown in cool regions with winter chill. They are rich in vitamin C and have wide culinary uses—for making drinks (including wine), desserts, jams, and jellies. Again, there are several different varieties and these will produce varying yields. Note that there may be restrictions on growing black currants in some areas; check with your local Cooperative Extension Office to find out more.

Black currants need to ripen on the bush for at least a week after their berries turn black. Don't pick them too soon, or they will be sour.

Soil, position, and cultivation

Black currants will flourish in most soils, but do best on those that are well-cultivated, rich, and well-drained. Ideally, they like a sunny position, but although they will tolerate partial shade, they do need to be in a sheltered spot which is protected from cold winds and away from frost pockets.

Dig lots of well-rotted compost into the ground a month or so before planting—which is best done in the fall or winter. Follow instructions for planting outlined opposite and space plants about 5 ft (1.5 m) apart. Cut all stems down to 2 in (5 cm) from the ground immediately and mulch around the roots.

Water plants well in dry periods and apply an annual mulch each spring. Keep the ground around the plants free of weeds by shallow hoeing. If necessary, control weeds by mulching, rather than digging, which would disturb the shallow root system. Because the roots are so shallow, a drip irrigation system is a good choice for black currants.

The number of plants to grow depends on the variety you choose, but an average-sized mature bush yields approximately 10 lbs (4.5 kg) of fruit a year.

Training and pruning

Black currants fruit on new wood each year, but as the new wood grows on old wood, you must prune regularly to maintain a good balance of new wood, while still keeping the plant bushy and allowing light to penetrate. After planting, cut out the weakest shoots at ground level—the remainder will fruit the following summer. From then on, prune the bush by removing a quarter to a third of the old wood each year. Take the oldest wood first—it will be very dark. Cut back this wood to one or two growth buds from the ground so new shoots grow.

Harvesting

Pick black currants when they are fully ripe—this is at least a week after they have turned from green to black. If clusters of fruit have both ripe and non-ripe fruit, the berries should be picked individually, which is a tedious task. Those picked on sprigs or trusses will keep marginally longer before use, but you should eat or freeze them fairly quickly.

Raising new plants

Take cuttings in the fall from healthy shoots of that year's growth. Cut off either end just beyond a bud so the cutting is about 10 in (25 cm) long. Push it into the soil. If the soil is very heavy, dig a trench about 6 in (15 cm) deep and sprinkle sand in the bottom. The cuttings should be about 6 in (15 cm) deep in the soil, with two growth buds showing above the surface, and there should be 9 in (25 cm) between cuttings. They will have rooted by the following year and may be transplanted to their permanent site.

Pests and diseases

Reversion virus and big bud are feared black currant disease in Europe; never import plants from Europe in case they carry diseases. Spider mites (see page 146) can distort and crinkle leaves. Armillaria (page 148) and gray mold (page 146) attack plants, especially in cool, damp conditions. Currants are alternate hosts for destructive white pine blister rust, so there are places where they cannot be grown.

SEASONAL GROWING GUIDE FOR BLACK CURRANTS

VARIETIES	PRUNING	HARVEST
Titania, Ben Sarek: Modern black currant varieties that are disease-resistant, compact, and produce heavy yields **Otelo** Midseason, old-fashioned variety. Heavy cropper that produces lots of sweet fruits **Ben Tirran, Ben Alder** Late-season, traditional black currant variety. Flowers late to escape frost	Pruning is the same for all varieties. Cut back the weakest shoots to ground level in the first fall. Thereafter, prune in late summer to fall, removing a quarter to a third of the old wood at the base of the bush each year to encourage new growth	Pick currants when they are ripe, in early to late summer. Pick individual currants if the whole truss does not ripen together

CULTIVATING BLACK CURRANTS

1 Plant canes in a sunny, sheltered position in fall, while the soil is still warm. Black currants may be sold in containers or as bare-root canes. Mulch well to retain moisture.

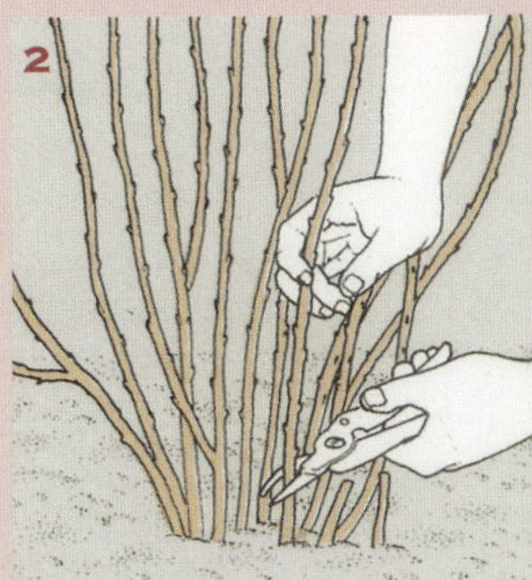

2 Cut back each shoot on a newly-planted bush to ground level, cutting just above an outward-pointing bud. The bushes will not fruit in their first year, but will grow strong and vigorous.

3 Take cuttings in fall. Rub all buds from the lower end and push the canes into a V-shaped trench. Leave to root before moving to their permanent site.

RED AND WHITE CURRANTS

Currants and gooseberries have been grown in Europe for years, but their popularity in North America waned years ago when it was discovered that they are an alternate host for white pine blister rust. There are still restrictions on currants in some areas; check with your local Cooperative Extension Office.

White currants are a variety of red currant and may be eaten fresh as a dessert. Red currants are more often used for jam and jelly-making, although they too can be eaten fresh. Both plants grow best in well-drained, but moisture-retentive soils. A sunny position, sheltered from cold winds, and where there is no danger of frost collecting to damage the early flowers, is the ideal site.

The average yield from an established bush is 4–5 lbs (1.8–2.2 kg). Some bushes will give twice this amount; cordons will yield less.

Planting and routine care

Dig plenty of well-rotted organic matter into the soil some weeks before planting and apply an all-purpose fertilizer with potassium. Buy two-year-old bushes and plant them as early as possible in the recommended planting period.

Drive stakes into the ground next to cordons and secure wires between these. Branches on bushes should be no more than 10 in (25 cm) above the ground.

Water in very dry weather and protect plants from birds. Pull out any suckers that appear from the roots or main stems. Control weeds by hand or by mulching. Apply annual winter mulches and feed with potassium fertilizer at the same time.

Pruning and training

After planting, cut back branches so each has only four shoots. The second winter, cut each branch back by half to an outward-facing bud and cut lateral shoots back to two buds. The aim is to produce an open-center bush that light can easily penetrate, and to produce spur

White currants have a delicate color and flavor; their glossy red relatives are more robust, and a great accompaniment to meats, such as roast lamb and beef.

CULTIVATING RED AND WHITE CURRANTS

1 Plant bushes in a sunny, sheltered spot. Stake them for support and prune each year.

2 As fruit begins to form, protect the crop from birds by covering bushes with netting. Water in dry weather.

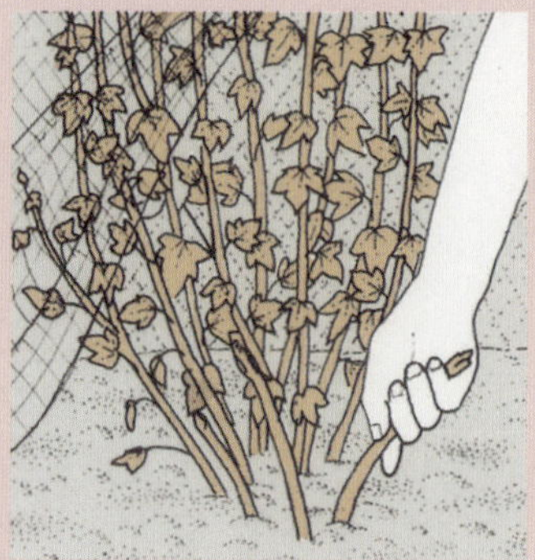

3 Pull out suckers as they appear near the roots of the main stems. Keep down weeds with gentle hoeing or pulling by hand.

wood that will bear the fruit. When established, cut the current year's growth back to about 1 in (2.5 cm) and cut out old wood to make room for new shoots. In summer, cut out lateral growths beyond five leaves.

Currants are trained and pruned in the same way as gooseberries (*see* page 138).

Harvesting

Pick the fruits as soon as they are ripe, harvesting whole trusses whenever possible. If they are not ripe, pick the currants individually.

Planting distances for red and white currants

Plant red and white currant bushes to the following suggested spacings. If more than one row of cordons is being grown, leave 4 ft (1.2 m) between them.

Bushes 5 ft (1.5 m)

Single cordons 15 in (40 cm)

Double cordons 2 ft 6 in (75 cm)

Treble cordons 3 ft (1 m)

BLACKBERRIES AND LOGANBERRIES

Native blackberries can be found in many areas, but cultivated blackberries are usually bigger, juicier, and tastier than their wild cousins. They may be trailing, semitrailing, or upright types, and some are thornless, to make harvesting fruit a painless process.

Loganberries are the result of an accidental cross between a blackberry and a raspberry in the late nineteenth century in California. The berries are dark red and may be eaten raw or used for cooking. They are trained, and propagated the same as blackberries.

Soil and position

Blackberries will grow in any soil, although those that are slightly acidic and rich in humus will produce the largest crops. They will grow just about anywhere, even in a north-facing site, and can be trained over a shed or along a fence. For the largest yields, choose an open, sunny position.

Planting and routine care

Plant blackberry canes as outlined on page 112, in late fall. Dig lots of well-rotted organic matter into the soil a month or so before planting. It is worth preparing the site well, because blackberries can remain in the same place for 10 years or more. Place canes about 6 ft (2 m) apart (slightly less for smaller canes; more for the more vigorous varieties) and stretch horizontal wires between them to provide support.

Water well in dry weather and keep the ground around the roots free of weeds. Mulch each spring with well-rotted manure or compost.

The number you grow depends on the varieties you choose—one cane of a vigorous variety can produce over 20 lbs (9 kg) of fruit.

Training and pruning

As the canes grow, tie them into the supporting wires, using one of the methods illustrated below. The one-year-old canes produce the fruit, so try to keep new and old wood separate. After harvesting, cut out all that year's fruiting canes at ground level. If necessary, reposition and

TRAINING BLACKBERRIES

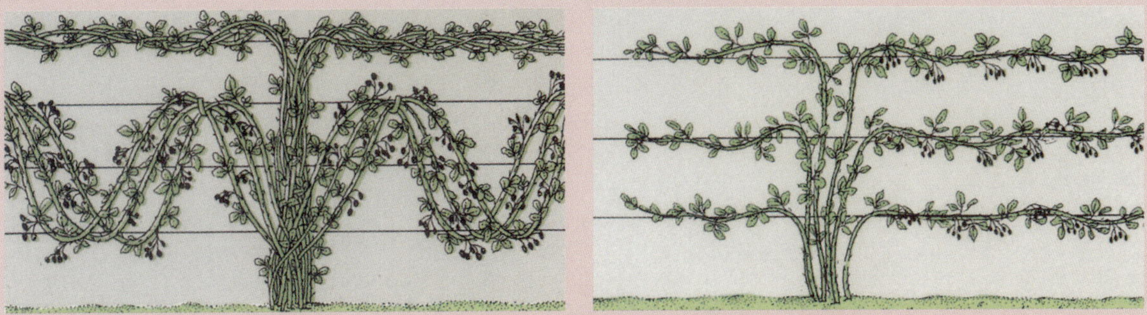

Training trailing blackberry
As the current season's canes grow, train them to weave in and out of rows of wires. As the new canes appear, train them to grow along the top wire.

Alternative sides
Train current season's canes to grow along one side of the wires, in a fan shape. Train the new canes, as they form, to grow along the other side.

SEASONAL GROWING GUIDE FOR BLACKBERRIES

Planting	**Late fall–early spring** Cut back each cane to 9 in (25 cm)
Training	**Summer** Keep tying in new growth throughout the growing season
Harvesting	**Summer–early fall** Pick berries when ripe and use them right away
Pruning	**Fall** Cut back fruiting canes to ground level

retie that year's new growth, which will fruit next year.

Harvesting

Blackberries ripen any time from mid-summer to fall, depending on the variety and, of course, the weather conditions. Pick them when they are ripe, handling them carefully so as not to crush them. Use them right away, as they will quickly turn soft and mushy.

Raising new plants

Bend over a new shoot in mid-summer, and bury the tip in 4–6 in (10–15 cm) of soil, holding it in place with a stone. Cut it away from the parent plant in early winter, when it has rooted, then replant it in its permanent site the following spring.

Pests and diseases

Check your canes regularly for signs of the following common pests and diseases: raspberry beetle (page 149), cane spot (page 147), and spur blight (page 147).

GOOSEBERRIES

The different varieties of gooseberry make it possible to produce crops right through the summer from late spring. As with currants, there may be growing restrictions. Check with your local Cooperative Extension.

Gooseberries are excellent for jams, chutneys, and desserts. Although usually grown on a bush, they can be grown on single, double, or treble cordons, which means they can be trained to grow against a wall in a small garden. They can even be trained to grow as a compact standard tree.

Soil and position

Gooseberries will grow in almost all soils, but prefer those which are well-drained but moist, and enriched by lots of well-rotted organic material. They like a sunny site, and should never be planted in a place where frost collects as they flower in early spring, and it will damage them.

Planting and routine care

Make sure the ground is completely free from all perennial weeds and apply an all-purpose fertilizer (preferably one that is rich in potassium) to the site before planting. Plant bare-root plants in the fall and winter, or plant container-grown plants any time it is not too hot. Leave 3–5 ft (1–1.5 m) between plants. The lowest branches should be about 9 in (25 cm) above the ground. Single cordons should be 1 ft (30 cm) apart, double cordons 2 ft (60 cm) apart, and triple cordons 3 ft (1 m) apart. They will need wire supports to train the branches.

Keep the ground free of weeds by shallow hoeing (in order not to disturb the roots). Water the plants in very dry weather and apply an annual spring mulch of well-rotted compost, together with a sprinkling of potassium-rich fertilizer.

Protect plants from birds by covering them completely with netting or by weaving black cotton thread around the branches. Tear out any shoots at ground level.

Mature bushes yield an average of 8 lbs (3.5 kg) of fruit each year.

Training and pruning

The aim with gooseberry bushes is to keep a cup-shaped form to the bush with a good open center. In the winter, cut summer growth shoots back by about half to two-thirds to a bud (leave the pruning until the buds start to swell if the bush has been attacked by birds—then you can be sure of pruning back to a healthy bud). If the bush is of a variety that has drooping growth, cut to an upward-pointing bud; cut upright varieties to an outward-pointing bud. Leave about eight branches well-spaced round the bush and cut the others to one bud from the base.

From then on, keep cutting the new growth back by half each winter and cut lateral shoots to about 3 in (7.5 cm) to encourage the formation of spurs. When the bush is really established, cut out old wood to maintain the shape and to allow light to penetrate evenly. Cut back leading shoots to about 1 in (2.5 cm) each year. Remove weak new shoots.

In summer, prune back sideshoots to about five leaves (do this on all types of gooseberry bushes, cordons included). This encourages the fruit buds to form.

Cordons can be created from cuttings. For a single cordon, cut off all shoots close to the main stem, leaving the strongest one. Tie it to a stake and cut it back by half the

Although green is the traditionally accepted color for gooseberries, many varieties are now available, including yellow and red fruits, too. Some varieties are too tart to eat fresh, so look for sweet or dessert gooseberries if you don't want to have to cook them before eating.

current season's growth each year. Shorten sideshoots to form fruiting spurs. Train and support double and triple cordons in a similar way but select two or three shoots initially. In a double cordon, train these horizontally and then cut off at an upward-pointing bud. In a triple cordon, train two shoots as for a double cordon and treat the center one like a single cordon.

Harvesting

If the crop is very heavy, thin the fruits, so they are about 1 in (2.5 cm) apart. If you like, you can thin them again to leave them 2–3 in (5–7.5 cm) apart, in which case you should get large gooseberries for the final harvest. Gooseberries for cooking do not have to be completely ripe before harvesting—those for eating fresh should be ripe or they will be tart.

Raising new plants

New plants can be raised from cuttings in the same way as black currants (*see* page 135), but they do not root as easily. Take cuttings in the fall, about 15 in (40 cm) long, from the current season's growth while there is still some leafy growth. Cut off the top 3 in (7.5 cm) of soft wood and rub out all the buds from the lower end, leaving three or four near the top. After that, treat the same as black currant cuttings.

Pests and diseases

Gooseberries may get powdery mildew (see page 145) and gooseberry sawfly (page 144). They are a host for white pine blister rust, and so cannot be grown in all areas.

PRUNING GOOSEBERRIES

1 In the winter, cut summer shoots back to a bud, always maintaining a good cup shape to the bush.

2 In the summer, thin the fruits by removing every other one. Prune back sideshoots to five leaves.

3 In winter, prune mature bushes to keep a cup shape. Cut leaders back by half and laterals to 3 in (7.5 cm).

DIRECTORY OF PESTS AND DISEASES

Preventing Pests and Diseases

It is inevitable that your fruit and vegetable crops will suffer attack from pests or succumb to disease at some point, but you can help to keep problems at bay with some common sense and regular maintenance. This is far easier than getting rid of a problem once it has taken hold.

Being strict and careful about garden hygiene is the most important first step in keeping your plants healthy. Tidy up plant debris regularly, so that dead or infected leaves, fruit, and twigs do not build up on the soil, infecting the other growing plants around them.

As you tidy up the fruit and vegetable beds, and whenever you water the plants, too, check your crops for any early signs of pest infestation or disease.

Always dig up and destroy plants affected by virus diseases to eradicate the problem quickly and weed your beds regularly to keep down weeds, where pests can live.

Choose plants wisely

Only buy plants, seeds, sets, or tubers from reputable suppliers and always look for varieties specified "certified disease-free." Choose disease-resistant varieties if you know that a particular disease is prevalent in your area. If a nursery or garden center looks messy, weedy, and dirty, choose to buy your plants elsewhere.

Don't cramp plants

Just as with people, plant diseases will spread more easily in crowded places, so always follow the recommended spacings when planting out. Plants that are crowded will also often be too cramped for air to circulate freely and this can lead to problems with rotting, fungal diseases, mildew, and more. They will also be competing with one another for water and nutrients and are likely to be weak and straggly as they race to grow up above their neighbors to reach the light. This makes them more vulnerable when pests and diseases do occur. What is more, it is harder for you to spot problems promptly if you cannot easily inspect plants because they are entwined with one another.

Beneficial garden creatures

All these creatures will prey on insect pests, helping to control them in your plot, though not eradicate them altogether.

- Centipedes
- Frogs, toads, and newts
- Soldier beetles
- Flower flies
- Lacewings
- Ladybugs
- Mealybug destroyer
- Rove beetles
- Tachinid flies
- Wasps

Greenhouse hygiene

It is particularly important to keep things scrupulously clean in the greenhouse, where the good growing conditions mean that pests and diseases will thrive just as readily as your plants.

Always scrub and disinfect pots and trays before and after use, clean up messes promptly, and clean the greenhouse itself thoroughly each

fall (*see* page 38). Open windows to allow air to circulate, and to discourage mold and fungus.

NATURE'S PEST CONTROL

Many insects and creatures survive on a diet of the pests that would like to eat your crops. If you can encourage them into your garden and fruit and vegetable beds, then they will do much of your pest control for you.

It may be easier to attract wildlife into a home garden than a vegetable plot, as the variety of plants is often greater there, but try to incorporate pollen-rich plants to attract flower flies and other insect predators that will help to control aphids. Ground beetles will attack the eggs of flies and caterpillars if you provide them with some shelter in the form of groundcover to protect them during the day.

Bottom left *Ladybugs are voracious aphid-eaters; attract them with flowering plants.*
Bottom center *If you have a pond, its frogs will prey on a variety of garden bugs.*
Bottom right *Beetles will prey on the eggs of many damaging pests, such as carrot fly.*

Treating Plant Problems

Vigilance and prompt action are the keys to successfully eradicating a pest or treating a disease. Inspect your crops regularly to spot signs of infection or infestation early.

Garden chemicals can be dangerous and should always be used as a last resort. Most insecticides will kill all insects, even the harmless or helpful ones, so only apply them to plants where a pest or disease is visible.

The following pages will help you to identify problems with your plants and suggest ways to treat or avoid them happening again, but it is a good idea to ask your local Cooperative Extension Office for advice on identifying and treating a pest or disease. They will be happy to recommend appropriate treatments or advise you on non-chemical methods where possible.

Using sprays safely

There will be times when the only practical solution is to use a chemical insecticide, fungicide, or other treatment. Always follow the instructions on the package carefully and only use a chemical for its intended purpose. Use a small sprayer, targeting your treatment rather than spraying a wide area. Do not spray chemicals on a windy day, when they could be blown off course or even into your face.

Take care when treating fungal diseases, as too much fungicide can be as harmful to many plants as the fungus itself. Always check the bottle for any specific warnings or test the product on one or two leaves before treating a whole plant or entire crop. Although it is best to confine insecticide spraying to only the affected plants, fungal diseases spread from spores, so treat the affected plant and those around it.

Be thorough and careful, and be prepared to apply the treatment again a few weeks later to catch pests that have hatched out of eggs since the first spraying. Wear protective equipment and clothing: long sleeves and pants, gloves, sturdy shoes, and goggles and a mask if you wish.

Never leave leftover chemicals in your sprayer—you might forget what they are and could use them for the wrong thing next time you need the sprayer. Spray any unused chemical onto a patch of gravel then clean the sprayer before you put it away.

Always store all garden chemicals on a high shelf in a locked shed. Keep a note of their expiration dates and take care to rotate your supplies so that you always use the oldest first. Take any half empty bottles you no longer need to your local recycling center and ask their advice about disposing of them safely.

Choosing insecticides

Systemic insecticides are absorbed by the plant and are used to treat sap-sucking insects, such as aphids and mealybugs, which consume the insecticide-laced sap and die. Contact insecticides are sprayed onto the surface of leaves and stems and kill insect pests that come into contact with them, killing pests that eat the leaves, such as caterpillars. Ready-mixed sprays are the most expensive option, but the easiest to apply.

VISIBLE PESTS

PEST	IDENTIFICATION	AFFECTED	TREATMENT
Aphids	Tiny insects, most prevalent in warm, dry spells, will settle in large numbers on leaves, shoots, and young fruit, causing distorted growth	Many crops	Spray with a blast of water from the hose to wash off the adults; apply insecticidal soap or recommended insecticide
Asparagus beetle	Orange-colored beetles and their gray grubs feed voraciously on the foliage, turning it patchy brown and distorted	Asparagus	Hand pick or spray off the plants to remove beetles. Cut foliage in fall to destroy the eggs
Blackfly or black aphids	Infestations of tiny blackflies, which cling to the stems and suck the sap, causing the plant to wither and growth to become distorted. Eventually the plant dies	Broad beans, but also pole and bush beans	Early crops usually escape, as the pods are harvested before blackflies are active. Pinch out growing tips of plants as soon as you notice blackfly and pick off and destroy any infected leaves. Spray with a jet of water from the hose to wash them off the plant
Greenhouse whiteflies	Tiny whiteflies and scalelike larvae collect on the underside of leaves of plants grown indoors. The leaves become marked with sooty mold and honeydew	Many greenhouse plants	Spray regularly with recommended insecticide
Mealybugs	Small pink insects gather in patches on stems and leaves. They are covered with a white woolly wax	Crops grown in greenhouse, citrus	Treat with the beneficial insect mealybug destroyer. Spray with insecticidal soap or horticultural oil. Dab with alcohol-soaked cotton swabs
Scale insects	Flat, scalelike insects, brown, yellow, or white. Found on the underside of leaves, clustered along leaf veins, and on stems	Citrus, bay, and other trees	Brush off the scales with a soft toothbrush, or dab with alcohol-soaked cotton swabs
Webworms, Tent caterpillars	Caterpillars of various moths that form large, tentlike structures in trees and shrubs	Many trees and shrubs	Control by removing and destroying the webs. Cut off affected branches and destroy. Use a recommended insecticide.

Other possible pests: **Cabbage caterpillers** (below), **Cutworms** page 147, **Weevils** page 147, **Flea beetles** (below)

LEAVES WITH HOLES

PEST	IDENTIFICATION	AFFECTED	TREATMENT
Bacterial canker	Disease spots appear on branches and ooze a sticky substance. These branches produce small, withered, discolored leaves and will eventually die back	Mostly plums, peaches, cherries, and tomatoes	Cut out all infected wood and seal cut surfaces with protective paint. Spray with recommended foliage spray in late summer or early fall
Cabbage caterpillars	Caterpillars eat through the leaves, making a mass of holes	Cole crops	Squash or pick off the clusters of yellow eggs and any caterpillars; use floating row covers
Flea beetles	Eats through seedlings in particular, leaving small, neat holes	Cole crops, radish, and turnips	Evening, rather than daytime, watering discourages the pest
Gooseberry sawflies	Leaves become stripped to their skeletons by caterpillars that are green with black spots	Gooseberries	Pick off eggs and caterpillars from the undersides of leaves. Spray with recommended insecticide in late spring or when symptoms first appear
Japanese beetles	Green- and copper-colored adults chew ragged holes in leaves and flowers. Grubs eat the roots of lawn grasses. Can arrive in hordes	Many, including apples, asparagus, and currants	Milky spore bacterium can kill grubs in lawn grasses. Adults may be controlled with pyrethrum, neem, or a recommended insecticide. Japanese beetle traps may attract more insects than they discourage.
Shothole	Leaves develop brown patches that turn into holes. Caused by leaf-spotting fungus or bacterial canker	Cherries, plums, and peaches	Feed plants every year, mulch in spring, and keep well-watered to guard against disease. Spray with a recommended fungicide in summer and fall

Other possible causes: **Slugs and snails** page 147

DISTORTED LEAVES

PEST	IDENTIFICATION	AFFECTED	TREATMENT
Artichoke curly dwarf	Large, dark spots appear on the leaves, and heads of artichokes are deformed	Artichokes	Dig up and destroy affected plants. Use disease-free stock or grow from seed
Celery fly	Maggots burrow into leaves, leaving trails of brown blisters. Growth becomes checked and plant will ultimately die	Celery, parsnips, and other root crops	Check leaves and crush blisters as soon as you see them. Pick off badly infected leaves and destroy them to prevent the pest spreading
Leaf roller	Larvae feed on new foliage growth, then curl the leaves around themselves to make a secure hiding place.	Many fruit trees	Beneficial insects like tachinid wasps will eat the larvae, as will birds. Pluck off affected leaves and destroy. Use a recommended insecticide.
Nematodes	Soil-dwelling microscopic pests attack roots and tubers	Many vegetable crops and strawberries	Plant resistant varieties and practice crop rotation
Onion white rot	Foliage becomes yellow and fall over; bulbs are soft, with fluffy white growth	Chives, garlic, onions, and shallots	Pull up and destroy infected plants. Onions grown from seed are less likely to be affected than those grown from sets. Follow a strict crop rotation plan and do not grow onions on ground you know to be infested for at least three years
Peach leaf curl	Attacks peach leaves, showing up first as large, red blisters that turn white, then brown. The leaves become curled and crumpled before dying and falling off	Peaches	Spray with recommended insecticide in late winter, as the buds begin to swell, repeating 10–14 days later and again in the fall, just before the leaves start to fall
Pear leaf blister mites	Tiny insects that feed on the leaves and produce tiny brownish pink blisters	Pears	Mites first appear in spring, so pick off and destroy infected leaves after this time. Bad attacks can be controlled with recommend insecticides
Powdery mildew	Patches of powdery white fungus cause leaves to be stunted and die	Many plants, including onion, strawberry, and grapes	Remove or prune out affected shoots and leaves and destroy them. Keep good air circulation around plants. Water and mulch plants well
Whiptail	The leaves begin to shrink towards the central vein; caused by nutrient deficiency	Cauliflowers, broccoli, and kohlrabi	Make sure the soil is not too acidic and, if so, treat accordingly
White blister	Leaves become covered with white blisters	Cole crops, salsify, and black salsify	Remove and destroy leaves. Caught in its early stages, this should not affect plants adversely

Other possible causes: **Slugs and snails** page 147, **Rust** page 146

Left *Cabbage caterpillars are bright and easy to spot: pick them off and destroy any clusters of eggs you find.*

Right *Aphids are among the most common pests on fruit and vegetable crops, and elsewhere in the ornamental garden. They will distort growth and ruin harvests. To reduce the problem and keep spraying to a minimum, try to encourage insect predators, like ladybugs, lacewings, and flower flies into your garden (see page 142).*

DISCOLORED LEAVES

PEST	IDENTIFICATION	AFFECTED	TREATMENT
Celery leaf spot	Brown spots with black centers appear on the leaves	Celery	Check that plants are free of disease before planting them in the garden. Spray with recommended fungicide
Downy mildew	Yellow spots appear on top surface of leaves and a furry, grayish-brown growth forms underneath	Cole crops, squash, and onions	Keep the atmosphere dry, avoid overhead watering. Remove and destroy infected plants
Fireblight	Causes flowers to blacken and shrivel, leaves to turn brown and wither, and shoots to die back	Apples and pears	Prune out infected branches and sterilize pruners and loppers after use. Plant resistant varieties
Greenhouse red spider mites	Tiny insects, which feed on the sap of plants, particularly tree fruits, causing the leaves to become mottled before turning a yellowish bronze and dying. They are active during the summer and early fall	Greenhouse plants	Spray with recommended insecticidal soap or horticultural oil. Keep leaves sprayed with water and dampen nearby paths and staging
Gray mold (Botrytis)	Brown spots appear on leaves, and gray mold then forms on these. Fruits begin to rot and are covered with gray mold. The plants will wilt and die	Strawberries, grapes, and many crops, especially in cool, damp conditions	Do not grow plants in cold, damp conditions; check greenhouse and cold frame plants regularly and provide good ventilation. Choose disease-resistant varieties if growing crops in winter
Leaf spots	Several different types of bacterial and fungal leaf spots cause brown spots, usually starting on oldest leaves first	Leeks, onions, beans, beets, and many others	Be sure to identify the cause of the leaf spot before spraying with a recommended fungicide. Provide good air circulation and destroy infected tissue
Mildew	The whole plant becomes covered in mildew and the leaves curl	Many	Spray with recommended fungicides and increase watering. Ensure good air circulation around the plants
Mosaic virus	Leaves and fruits of plants turn yellow and mottled. Eventually growth becomes stunted and the fruits do not develop properly	Many, but particularly cucumber	Disease is spread by flying insects, so control these by covering susceptible crops with floating row covers. Plant resistant varieties
Red spider mites	Tiny creatures that feed on the leaves, turning them a dusty red color. Spin very fine webs between the leaves	Beans, peas, and fruit trees	These usually only attack plants in a very hot, dry atmosphere. Water regularly and mulch, to keep up the moisture levels
Rust	Brownish spots on leaves, that become twisted or distorted	Beans, perpetual spinach, leeks, and other plants	Pull off leaves and destroy them. Spray with recommended fungicide
Virus diseases	Leaves become discolored and often distorted	Many	Dig up and destroy affected plants. Grow only certified disease-free plants to minimize the risk of disease

Other possible causes: **Bacterial canker** page 144, **Carrot flies** page 151, **Leaf miners** page 148, **Powdery mildew** page 145

Far right *Red spider mite attacks peas and beans; the greenhouse red spider mite is a common problem on all kinds of plants grown indoors.*

Right *Rusty-looking brown spots on leaves are a sign of rust. Remove and destroy any affected leaves to halt the spread of the problem.*

DISTORTED OR WILTED SHOOTS

PEST	IDENTIFICATION	AFFECTED	TREATMENT
Fusarium wilt	A fungal disease that lives in the soil and causes the water-conductive tissues of the plants to collapse	Tomatoes	Improve air circulation. Practice crop rotation. Remove and destroy infected plants
Slugs and snails	Irregular shaped holes in leaves, slime trails on surrounding paths and stones, complete destruction of young plants	Many, particularly young seedlings	Remove plant debris regularly to avoid creating conditions where slugs and snails thrive. Pick off slugs and snails and destroy them and try to encourage wildlife into your garden to control these pests for you. Use slug pellets only where they cannot cause harm to other wildlife or animals
Weevils	Chewing insects whose adult form chews leaves, blossoms, and fruit. Larvae feed on roots, causing plants to wilt and die	Many, especially strawberries	Encourage beneficial insects and birds. Hand pick adults at night. Spray with recommended insecticide

Other possible causes: **Pear leaf blister mites** page 145

PROBLEMS WITH STEMS, BARK, OR BRANCHES

PEST	IDENTIFICATION	AFFECTED	TREATMENT
Bacterial canker	Long, gummy lesions develop on branches. If left, they get larger until the whole branch dies	Cherries, plums, peaches, and other pitted fruits	Cut out diseased wood and destroy it. Practice good garden hygiene, cleaning up fallen leaves and fruit. Select resistant varieties
Cane spot	Fungal purple spots appear on the canes. These later change to white and the cane eventually splits	Blackberries, raspberries, and loganberries	Cut out and burn badly infected canes and spray with recommended fungicide
Cutworms	Worms eat through the stems at ground level. The greenish-gray caterpillars may be visible on the ground around the plants	Many vegetable seedlings	Make sure weed growth is controlled, to make it hard for the pest to move from plant to plant. Hand pick pests. Encourage birds
Nectria canker	A mass of red spots develops, usually on old wood	Tree fruits and currants	Cut back all affected areas to at least 6 in (15 cm) past the infection and destroy clippings. Seal all pruning cuts
Spur blight	A fungus disease that, like cane spot, causes purple spots on canes. The leaves wither and canes snap in bad cases	Blackberries, raspberries, and loganberries	Cut out and destroy infected canes and spray others with recommended fungicide
Woolly aphids	Aphids leave tufts of waxy wool on branches and twigs. Some may develop into galls	Apples and pears	Some varieties are resistant. Use biological controls. Spray with recommended dormant spray

Other possible causes: **Fireblight** page 146

Far left *Apple trees are particularly at risk from woolly aphids, which can lead to more serious disease.*

Left *Shredded leaves—even plants that disappear overnight—are often the result of slug attack. These slimy pests can destroy an entire crop.*

WILTING OR WITHERED PLANT

PEST	IDENTIFICATION	AFFECTED	TREATMENT
Armillaria root rot	Leaves wither, shoots die back, and fungal growths appear. The plant will eventually die	Vines, figs, and other plants	Dig up and destroy dead plants and change or sterilize soil before attempting to replant in the same place
Blackleg	A black rot develops at the base of stems, causing them to become soft; leaves turn yellow and the plant eventually dies	Potatoes	Use only certified disease-free seed potatoes
Cabbage root maggot	Small white maggots feed on roots (particularly on recently transplanted seedlings), causing leaves to turn a blue-gray color; plants wilt and collapse	Cabbages and other cole crops	The cabbage root fly lays its eggs on the soil surface and the hatched maggots burrow into the ground, eating stems and roots. Prevent them doing so by fitting a plastic or rubber disk around the stem of the young plant and pushing it just below the surface to create a physical barrier
Clubroot	One of the most common diseases of cole crops. Unpleasant-smelling swellings form on the root, stunting the plant's growth. The leaves turn yellow and wilt	Cole crops, turnips, and radishes	Always dig, rather than pull, roots up to ensure that none are left in the ground. The disease can stay in the soil for years. Diseased plants must be discarded. Lime the soil. Practice crop rotation. Grow crops in raised beds on poorly drained soil to reduce attacks and look for disease-resistant varieties
Crane flies	Fat, gray-brown grubs living in the soil attack the roots, causing plants to turn yellow and wilt. The plants may eventually die	Some vegetables, particularly cole crops	Fork the soil to turn it over and expose the grubs to birds. Alternatively, spread grass clippings on the soil, cover with cardboard overnight to draw the grubs to the surface then uncover and remove the now-infested grass clippings or leave the grubs for birds to pick out. Leatherjackets are crane fly larvae, so look out for adult flies and discourage or get rid of them
Damping off	A fungus causes seedlings to rot at ground level and die	Seedlings indoors or grown in crowded, wet conditions. Plants only vulnerable until they develop their third pair of leaves	Destroy infected seedlings. Sow sparingly, water carefully, and ensure good ventilation to avoid the disease. Always use clean flats and cell packs, and sterilized potting mixture. Only water with clean water, not rainwater
Dieback	Fungus that makes young branches and shoots wither and die suddenly	Mostly fruit trees	Prevent the fungus entering breaks or cuts in the wood by sealing any pruning cuts with wound-sealing paint. Cut off affected branches and seal the cuts
Leaf miners	The larvae of moths, flies, and beetles tunnel through plant leaves, leaving ragged trails and holes	Leafy crops, including Swiss chard and spinach	Use floating row covers to prevent flying insects laying eggs. Remove infested leaves
Onion flies	Maggots hatch from eggs laid on the surface of the soil and tunnel into the developing onion. It becomes soft and unpleasant. Attack can be detected by the purple-gray streaks that appear on the leaves	Onions	Onion sets are not usually attacked, and the pest is most prevalent on dry soils. Dig up and destroy infected plants. Attacks are less likely if the plant is handled as little as possible during development. Use floating row covers to keep flies of the growing crop
Stem rot	Roots shrivel, causing foliage to turn yellow and stems to turn a browny red color before rotting	Beans, peas, and tomatoes	Follow strict crop rotation and avoid replanting on infected ground. Always use sterilized potting soil for plants grown in containers
Violet root rot	Foliage turns yellow and the roots become covered with a purple-colored fungus. Although the attack usually starts on older plants, the disease will quickly spread to other crops	Mostly carrots, parsnips, and asparagus, plus some fruit crops	Dig up affected plants and dispose of them
Wirestem	The stems of young plants turn brown and begin to shrink	Cole crops	Raise seedlings in sterilized potting soil and avoid overwatering

FRUIT DISORDERS

PEST	IDENTIFICATION	AFFECTED	TREATMENT
Apple mildew	New spring shoots and leaves are covered with a powdery, gray-white coating	Apples	Cut out and destroy badly affected growth. Remove all diseased shoots the following fall. Make sure the tree is well watered in dry weather. Spray with recommended fungicides to help keep the disease at bay
Apple sawflies	Tiny antlike flies lay eggs in spring blossoms. Caterpillars hatch and burrow into the developing fruit, causing scarring on the skin and the fruit to fall before it is ripe	Apples and plums	Spray with recommended insecticide weekly when the blossoms start to fall
Apple scab	Green, brown, or black scabby blisters form on the skin of the apple, often making it distorted. Leaves and shoots can be similarly affected. Blisters on leaves and shoots will burst through the bark, causing cracks and scabs that can let in other infections	Apples	Plant resistant varieties. Remove and destroy all infected growth. Spray the tree with a recommended fungicide as soon as the flower buds appear
Bitter pit and cork spot	Brown spots develop on apples, running through the flesh	Apples	Mulch to conserve moisture. Spray with recommended calcium sprays
Brown rot	Soft, fuzzy brown spots develop on fruit and then soon grow, sometimes enveloping the whole fruit	Many pitted fruits, including peaches and nectarines	Pick off infected fruit and prune out spurs, which may also be infected. Plant resistant varieties. Spray with a recommended fungicide
Codling moths	Caterpillars burrow into the fruit of apples and pears. You may spot entry holes through the eye end of the fruit, but damage is rarely spotted before the fruit is cut open	Apples and pears	Practice good garden hygiene; clean up fallen fruit and leaves. Wrap trunks with sticky barriers. Spray with recommended insecticide
Gooseberry sawfly	Caterpillars (green with black spots) eat the leaves, reducing them to skeletons	Gooseberries and currants	From late spring onwards pick off and destroy any caterpillars. Spray with recommended insecticide
Pear midges	Only occurs in certain places, but once it has attacked a tree, it will continue to do so until the tree has been treated. The maggots of the pest bore through the skin and feed on the pears, slowing down the growth and making the fruit distorted. Infected pears will soon fall off the tree	Pears	Remove infected fruit and destroy it. Keep the ground beneath the tree well-cultivated, so that predators have a chance to eat the pests, before they climb up the tree
Plum curculio	A type of weevil, curculios have two stages of growth. Adults chew on leaves and fruit, and the females lay their eggs in the fruit. Larvae feed in the fruit and then drop to the soil below	Plums	Practice good garden hygiene, clearing up dropped fruit and leaves in fall. Shake trees to dislodge pests onto a tarp, then destroy them. Spray with recommended insecticide
Raspberry beetles	Maggots tunnel into the fruits, making them distorted and inedible	Raspberries	Spray with recommended insecticide as the fruits begin to change color
Silverleaf	Leaves turn a silvery color and the inner wood of branches turns purplish-brown. Eventually the branches will die	Many, including plums	Cut out all infected wood to about 6 in (15 cm) beyond the discolored area and seal the wounds. Make sure all pruning cuts are well sealed, because the disease enters through cut or open spots on the wood
Stink bugs and tarnished bugs	Brightly colored stink bugs are winged insects about 0.4 in (1 cm). Tarnished bugs are a bit smaller, brown with a V marking on their shieldlike backs. They chew leaves and stems	Many fruits, including berries, pears, and plums	Cover fruit with floating row covers to exclude the pests. Some beneficial insects may help, including tachinid flies. Apply recommended insecticide
Strawberry weevils	Fruit is eaten as it ripens, as if by birds, but beetles up to 0.4 in (1 cm) long ,with long snouts, can be found beneath the plants	Strawberries	Keep the ground around plants free of plant debris and weeds to reduce the places for beetles to hide. Spray with recommended insecticide when plants are not in bloom

Other possible causes: **Many** page 145 and 147, **Dieback** page 148

PROBLEMS WITH BEANS, PEAS, TOMATOES, AND LEAFY VEGETABLES

PEST	IDENTIFICATION	AFFECTED	TREATMENT
Anthracnose of beans	Pods and stems develop sunken dark brown patches and brown spots develop on leaves	Mostly pole and bush; sometimes lima and green beans	Dig up and destroy any infected plants and do not save seed from them for replanting next year. Follow a strict crop rotation plan to avoid growing beans on the same site the following year
Cucumber beetles	Oval, greenish-yellow beetles with black spots or stripes. Chew on plants and also transmit bacterial wilt and other diseases	Cucumbers, beans, squash, and melons	Cover crops with floating row covers. Pull up and discard infected plants. Spray with recommended insecticides
Blight	Brown-black blotches form on the upper side of leaves and a white furry coating underneath. Leaves eventually turn brown and rot. Fruits also develop a brown rot	Beans and peas	Avoid growing in the damp conditions that encourage blight. Avoid overhead watering. Cut off all affected foliage. Buy certified disease-free seed
Blossom end rot	Brown or black circular discolorations appear on the bottom of the young, developing fruits	Tomatoes, peppers, and eggplants	Keep plants evenly moist at all times. Pay particular attention to plants grown in growing bags, which can dry out quickly and may need watering several times a day
Poor ripening	Hard green or yellow patches appear on fruits, particularly on lower trusses	Tomatoes	Keep plants evenly moist and the temperature not too high. Try not to grow plants in poor soil or to let them grow too fast in poor light
Colorado potato beetle	Black-and-yellow striped beetles lay yellow shiny eggs and have brownish-red larvae	Potatoes, tomatoes, and eggplant	Handpick and destroy adults and eggs. Spray with recommended insecticide
Celery heart rot	Centers of celery clumps are wet, with a slimy brown rot	Celery	Slug damage or frost damage allow bacteria to enter plant, so protect from both to minimize attack. Follow a strict crop rotation plan to avoid growing celery again on infected soil
Cabbage whiteflies	Small white flies, which feed on the underside of leaves, leaving sticky discoloration. The growth of leaves is held back	Cabbage and other cole crops	Spray with recommended insecticide
Mexican bean beetles	Resemble ladybugs but with orange-brown coloring and eight spots. Adults and larvae eat all parts of the plant	Beans	Handpick and destroy adults and eggs. Spray with recommended insecticide
Gray mold	The upper surfaces of the leaves display yellow spots and mold appears on the under surfaces	Tomatoes, beans, lettuce, and many other crops	A warm, humid atmosphere is the perfect situation for this disease to develop and spread. Maintain good ventilation and control the disease with copper-based sprays. Grow disease-resistant varieties. Remove and destroy affected foliage promptly
Pea and bean weevils	Small, light brown beetle, which is active at night when it eats leaves. Only affects young plants	Peas and beans	Hoe between rows to disturb sleeping insects and make them good prey for birds
Squash bugs	Adults are dark brown or black; nymphs are smaller and pale green. Both suck stems, leaves, and fruits	Squash and pumpkins	Use floating row covers. Handpick and destroy adults and eggs. Spray with recommended insecticide
Thrips	Tiny, winged insects that eat leaves. If not controlled, they can cause stunted growth	Lettuce and other leafy greens	Keep plants well-watered and spray them well, too. Encourage beneficial insects, such as lacewings and minute pirate bugs
Splitting	The skins of the developing fruit splits either down or around the fruit	Tomatoes	Water and feed the plant regularly. Maintain even moisture in the soil at all times—do not allow the plant to dry out and then water it generously
Virus	Virus cause distorted leaves and stunted fruit. Tobacco mosaic virus (TMV) can be very damaging, seriously reducing crops	Tomatoes	Destroy affected plants and sterilize tools. Plant resistant varieties

ROOT VEGETABLE DISORDERS

PEST	IDENTIFICATION	AFFECTED	TREATMENT
Beet fly	Maggots hatch on beet leaves, tunneling through them	Beets	Check leaves frequently and pick off and destroy damaged ones
Curly top	A viral disease that causes foliage to become distorted, thick, and leathery	Beets, carrots, and many other vegetables	Spread by leafhopper insects, so use floating row covers to keep pests off the crops
Black slugs	Slugs eat into developing roots	Many, including root crops	Slugs are hard to control, as they live beneath the ground, so are difficult to detect until you harvest the crop and discover the damage. Try setting traps, as described for wireworms (below)
Carrot fly	Maggots eat into roots. Leaves turn a reddish color then yellow, and wilt	Carrots, parsnips, celery, and parsley	Sow seed as thinly as possible to keep later thinning to a minimum. Avoid handling plants and any thinnings as much as possible, as this releases a scent that attracts the flies. Sow early and late to avoid the time when adult flies are most active. Try erecting a barrier around the crop, growing in pots more than 18 in (45cm) high, as this is above the level at which the ad ult flies fly, or growing under row covers. Companion planting with onions or dill is also said to deter the adult flies
Downy mildew	Foliage is the first to be affected, wilting and collapsing. A white mold develops on bulbs and they do not store	Onions	The fungus lives in the soil for many years. Practice crop rotation or replace infected soil. Dig up and destroy affected crops
Onion white rot	Roots and bulbs develop white, fluffy growths and eventually rot. Foliage topples	Onions	Sow disease-resistant varieties and follow strict crop rotation plans. Discourage carrot flies, as these help to spread the disease
Potato blight	Occurs in damp, fall conditions, so early varieties should not be affected. Leaves become blotched with brown marks and the lower ones turn yellow. The tubers develop similar brown marks and turn rotten	Potatoes	Spray with recommended fungicide
Potato cyst nematodes	Small cysts develop on roots. The crop is vastly reduced and plants may wilt and die. Earlies may escape attack as the pest is not usually active until midsummer	Potatoes	Use certified disease-free seed potatoes and follow a strict crop rotation plan
Potato scab	Occurs in alkaline soil or soil that does not contain sufficient organic matter. Scabby patches occur on the skin of the potatoes	Potatoes, but also found on beets, radishes, and turnips	If caught in the early stages, scabs can be removed before cooking. Later, too much of the crop will be affected. Dig up and destroy if the attack is severe
Swift moths	Whitish caterpillars feed on plant roots. The caterpillars may be visible when digging or hoeing the soil	Carrots, parsnips, and endive	Pay close attention to garden hygiene. Pick out and destroy any caterpillars you see in the soil
Sclerotina	Plant wilt suddenly, and stems become brown and rotten. White mold and hard, black sclerotia fall into the soil	Bulbs, carrots, and parsnips	Dig up and burn infected plants. Follow a strict crop rotation
White rot	Attacks onions and leeks, turning the leaves yellow. They wither and die, revealing the base of the plant, which will be covered with a white mold	Onions, leeks, and garlic	Dig up and burn infected plants. Follow a strict crop rotation and sow seed very thinly. Seed drills can be treated with a recommended fungicide to help to deter disease
Wireworms	Larvae of click beetles eat into developing roots. The small wormlike creatures may be seen in abundance around the roots as you dig or hoe	Root crops, potatoes, and lettuce	Make traps by sticking a short length of stick into a potato and pushing it into the ground. The larvae will eat into this. Pull it out and destroy it

Possible other causes: **Cabbage root maggots** page 148, **Leatherjackets** page 148, **Clubroot** page 148

KEEPING ANIMALS

Keeping Animals
Pigs
Goats
Chickens
Ducks, Geese, and Turkeys
Bees

Keeping Animals

Livestock can be a delight that provides you with highly nutritious and tasty produce. Although, it can cost you more money than any you might make or save. Try to remain businesslike—they are not your pets.

Keeping chickens is a good place to start if you are not used to livestock. With a secure coop, they need little other attention.

The first and most important thing to remember is that keeping any livestock is a responsibility. The happiness and welfare of the animals is in your hands and even if you eventually mean to put them on the table in a tasty casserole, you have a duty to give them a happy existence while they are alive. Although many backyard homesteaders believe they are saving their animals from factory-farm conditions, they may inadvertently be inflicting just as much cruelty by not understanding the animals' needs.

Food and shelter

Always ensure that you have housing ready before your birds or animals arrive. You don't have to buy expensive goat huts or chicken coops: if you already have outbuildings or sheds, adapt and use these. Your animals will be happiest if you can keep them in conditions that emulate their natural life as closely as possible; and happy animals are healthy animals.

It is difficult to give rules about feeding, because animals have likes and dislikes and differences in appetite, just like people. Spend time watching your animals and get to know what they do and do not like.

Keeping healthy

This book does not aim to offer advice on treating animals that are ill—you will always need professional help from a vet—but animals kept on a small scale in clean, healthy surroundings should seldom succumb to disease. The rule is that if you notice anything wrong with your livestock isolate them to stop the problem spreading to the others and call the vet.

Balancing the books

Few people would claim that their eggs or honey cost them nothing, or that their pork was vastly cheaper than any they could buy. The point about home-produced food is that it is of an infinitely higher quality than you'll find in the supermarket.

It is a good idea to keep records—how much honey the bees produced and when; the amount of milk your goats give through the year; the eggs you get each day, and so on. By doing so you can spot problems and gain a more pragmatic view of your enterprise.

Start small

Do not try to become a poultry-keeper, apiarist, goatherd, and hog farmer all in one week. Do it slowly, one type of animal at a time, making sure you can fit them into your daily routine. If you try to do everything at once, you will be far more likely to fail and to give up.

Before keeping any of the animals on the following pages, you should check with your local authorities to ensure you are not contravening any bylaws or ordinances and to get all the necessary licenses.

Keeping Pigs

If kept in the right conditions, pigs are fairly easy animals to keep and rear, but they demand daily hard work and are big, strong animals, so be prepared for the physical aspects of raising swine.

Mucking out the pig pen is at least a daily chore and heavy work. The pigs must be moved into the yard while you do it and may not always be cooperative when you try to guide them outside. An adult pig is strong enough to bowl you over.

When you raise pigs, you feed them to a specific weight for slaughter, and then direct the processing. If you plan to make money from your pigs, remember that a sow will have eight or 10 piglets each time, so you are likely to have some to sell. To get the best price for your butchered stock you may need to have the meat processed into bacon, sausages, or other goods that can be sold at a greater profit.

LIVING REQUIREMENTS

There are two main ways in which you can keep pigs: in a permanent yard with a sty or shed, or to allow them to live free-range in a fenced pen—at least 150 square feet (15 square meters) per pig.

Never keep your pigs in a sty with a concrete floor. You may find this option suggested in old books, but pigs live by the snout—that is, their whole life consists of rooting through the ground—and most people would now agree that to deny them this by keeping them on concrete is cruel.

Licenses and regulations

Before you can buy any pigs—even just one pig—you must check with your local city or county zoning office to see if there are any ordinances prohibiting swine.

Next, plan ahead if you are planning to breed your pigs for meat. Make sure you have a butcher lined up for when you plan to have your pig slaughtered. Processors and butchers have busy schedules and are often booked up for months in advance.

Space and fencing

To let pigs roam freely demands a lot of land—at least 6 acres (2.4

It is not a myth that pigs love to wallow in mud. Wherever you keep your pigs, you must allow them a muddy, wet spot where they can cool down in hot weather.

hectares)—but it is ideal for both you and the pigs. Pigs are the best cultivators of all; better by far than any power equipment that is subject to breakdown. They will hunt out all the roots—weeds and all—turning the ground over and fertilizing it at the same time. You can be sure that yours will be the most delicious pork you have ever tasted if it is kept in this way.

Pigs will lean against the fences that confine them and rub against posts and rails, soon dislodging the stakes and weakening the whole structure, unless it is really strong and firm. Fencing should be hog panels or electric wire. At the bottom there should be at least one strand of barbed or electric wire to stop them nosing their way through underneath. The fence should be at least 3 ft (1 m) high; pigs can jump, especially if they are spooked or surprised. If your pigs escape (and they almost certainly will from time to time), they are easy to retrieve. Providing they know you and you have a bucket of their favorite food, they will follow you back to the yard.

Providing shelter

Housing for pigs can be as simple or as sophisticated as you wish to make it, as long as it is dry and warm. Dirt floors are fine, as pigs love mud. An alternative is a concrete floor, covered with wood—concrete on its own is too cold. In either case, a layer of straw as bedding will be appreciated.

Some homesteaders fatten their pigs in small "finishing units," a small building with a wooden slatted pen adjacent to it. Ideally you should be able to shut the pigs in, but if one side of the shed is open, make sure it is not in the direction of the prevailing wind or driving rain.

Pigs rarely soil their living quarters during the day, preferring to do this outside, so the house should not get too mucky, but this area should still be cleaned out daily, because if they are shut in at night they will muck somewhere. It will make your job of cleaning out easier if the accommodation is tall enough for you to stand up inside.

Somewhere to wallow

Pigs really do love mud and, in hot weather in particular, it is important to provide them with a wet, muddy hollow where they can wallow. The mud will help them to cool down and when it dries on their skin it will, like sunscreen, also help to protect them from burning in the sun.

LIVING REQUIREMENTS

Your decision about how to keep your pigs could be made for you by the amount of land you have. To keep pigs free-range you will need a large amount of fallow land that you can divide into areas for the pigs. They will need to move to a new area whenever each one is cleared, but cannot return to a patch of land for at least three years. In a permanent pen, you will need to provide food for the animals and you must provide water troughs in both situations.

OPTION	HOUSING	PROS	CONS	FEEDING AND OTHER REQUIREMENTS
Permanent yard	Yard surrounded by strong wall or fence at least 3 ft (1 m) high of strong hog wire fixed to sturdy posts. Make the yard large enough for pigs not to be overcrowded. A shed or other shelter is needed, with deep bedding	Easily contained to stop pigs from eating your crops. Strong barriers stop pigs escaping. Convenient for farrowing sows and looking after piglets	Barriers must be very secure, as pigs can easily knock them over	Throw dried straw into the yard from time to time for pigs to root in. Feed with garden waste, potatoes, other root vegetables, protein supplements, and vitamins
Free-range	Area marked out with very strong fencing. Movable pig shelter made from bales of straw with a corrugated iron roof. Provide deep bedding and encourage pigs in at night	Outstanding-tasting meat; pigs will cultivate rough ground for you	Do not move pigs back to same patch for at least three years: parasites that attack them will remain in the soil for this long	About 10 pigs can be kept in 0.5 acres (0.2 hectares); move them to another patch when they have exhausted each one. If grass is plentiful, feed once a day with meal or concentrated food. Supplement (above) as necessary

ACQUIRING PIGS

As with all livestock, it is best to acquire your pigs from a local farmer or dealer, so the breed you buy will depend on what is available. Livestock markets and auctions are busy and intimidating places and not for the faint-hearted. A local breeder may be a better option for a small-scale homesteader.

Some of the most common breeds are shown on page 159. Bear in mind that some of the new breeds, developed specifically with meat production in mind, could be more skittish and susceptible to disease than some of the old-fashioned types. Some traditional breeds of swine are even in danger of dying out as they are supplanted by modern breeds. Keeping your own pigs is also a good chance to try some of these older breeds, such as the Tamworth or the Hereford.

The Duroc is a good and popular breed, as are the Hampshire and the Yorkshire. These three are the most popular breeds in the United States. Crossbreeds are also good pigs to keep, and all these breeds are renowned for being good-tempered.

Buying feeders or breeding pigs

There are two possible systems for small-time pig-keepers to adopt. One is to buy eight-week-old feeders, keep them for 10–12 weeks (or longer if you prefer), and then slaughter them, and the other is to keep a sow and breed from her.

The first method is undoubtedly the most economical and also the least trouble. It also means you can keep just a couple of pigs to eat yourself, and you can sell the rest. If your sow has 8–10 piglets in a litter, you will need plenty of space to keep them comfortably until you sell them.

When buying feeders, look for piglets with long, lean backs and no signs of lameness. You will need a strong cage to get them home and a pickup truck or flatbed trailer to transport the cage.

Farrowing rails

These are essential at breeding times to prevent the sow from lying on her piglets. You can construct a frame by bolting or screwing wooden rails to some low, sturdy stakes. The piglets will naturally choose to huddle in the corners of the sty and will be protected by the rails when the sow lies there with them.

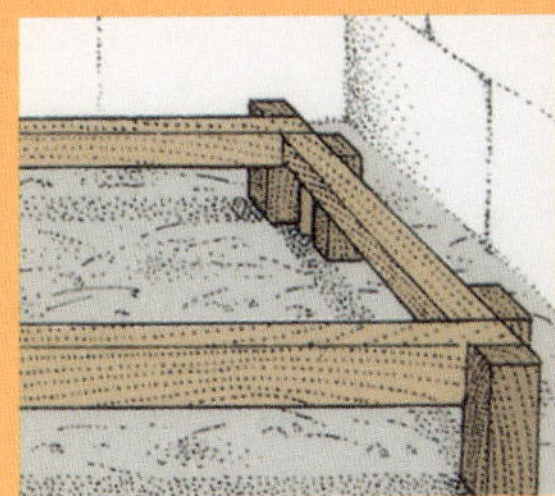

Moving pigs

Getting a pig to go where you want it to can be a challenge. A simple, but traditional aid is a long sheet of plywood, with hand holes cut in

BUILDING A PIGSTY

You can build a sty for your pigs if you wish to improve their living conditions. This can easily be constructed from cement blocks with a corrugated iron roof. Make it around 8 x 8 ft (2.4 x 2.4 m) and tall enough for you to stand up inside when cleaning it out. If you want to include windows, place them near the roof of the sty, opening outwards from the bottom, so that the pigs have ventilation without a draught. A yard is easy to clean out and essential for giving the pigs somewhere to exercise and explore. Always include a sturdy drinking trough; pigs drink a lot of water, although they do not mind if it is not clean.

the top edge. Use this to direct your pigs in and out of a trailer and to and from the yard or pigsty when you get them home.

FEEDING

You are unlikely to be able to grow sufficient quantities of the root vegetables and potatoes on which pigs live, so you will almost certainly have to feed them farm grains too.

Even if you do have a plentiful supply of root vegetables, if you are aiming to fatten pigs for slaughter, supplementary feeding is essential. The best feed is farm grains, including corn, barley, oats, and wheat. Swine also need supplementary protein, minerals, and vitamins. Eight-week-old feeders will need about 2 lbs (900 g) of feed per day, given in a morning and evening feed. This amount should gradually be increased to 4–4½ lbs (1.8–2 kg) by the time they are about four months old. Many pig owners also add an antimicrobial additive to their pig's feed.

A sow being used for breeding will want more and then still more food while she is pregnant and feeding her piglets. In addition, she will need milk (preferably from goats if you keep them).

Although pigs naturally root about for food, it is a good idea to feed them in a trough. This will save it from being wasted by getting trampled into the mud or urinated on. Scatter vegetables and other larger foods (below) across their yard to give them something tasty to hunt around for.

Pigs will root for food on the ground, but feed and small food are best given in a trough or bucket, so they don't get trampled underfoot and wasted.

Will pigs eat anything?

Besides their feed, pigs will enjoy a variety of other foodstuffs. Root vegetables are very important—Jerusalem artichokes, rutabagas, turnips, parsnips, and so on—and these can be grown in their yard or on a patch of land where they are allowed to roam, so that they can dig them out themselves. They like potatoes (although these are more digestible if boiled), and will appreciate some greens. Other things they like are apples (windfalls), clover, and any plants from the cabbage family. You can also give them peelings or trimmings from vegetables and they will happily consume any bruised or damaged fruit or vegetables that you do not want to keep for yourself.

If your pigs are free-range and eating pasture, they will really only need the supplementary feed—the rest they should be able to forage for themselves.

Pigs drink lots of water, but unlike other animals such as goats, this does not have to be sparkling clean. Give the water in troughs and make sure there is always enough in them.

MANAGEMENT AND BREEDING

Your pigs should give you remarkably little trouble and generally need very little attention.

In hot weather, throw a bucket of water over the pigs and another onto the ground in their pen. Pigs dig holes to sleep in and like these to be damp. Make sure your pigs have somewhere to shelter from hot sun, as they can suffer from sunburn.

If you do want to breed from a sow, let her have one season first (this will be at about six months old), and then breed from her when she is about a year old. It is possible to artificially inseminate pigs, although the timing is critical, and you must be sure of what you are doing, too.

The alternative is to take her to a local pig-breeder, providing they are willing to take her (some will not, because of the risk of infection). They will generally leave her with a hog for a few days.

Taking care of piglets

The gestation period for pigs is 14–16 weeks, and a pig will usually have between 8–10 piglets, which she can rear easily without any additional help. Shut her in the shed or shelter to have her litter. It is always a good idea to construct a farrowing rail (*see* page 157) in the shed where the sow has her litter, to stop her lying on her piglets and inadvertently crushing them.

A farrowing rail is a set of very strong bars, which should be raised about 9 in (25 cm) from the ground and a similar distance away from the wall. It must be strong—old iron bed rails are ideal, or use sturdy wooden posts and rails—and should be firmly secured to the wall, or the pig will merely get her snout underneath and wrench it free.

The farrowing rails should be sited along the two walls of the sow's favorite corner of the shed, which is where she is most likely to have her piglets. It is a good idea, too, to stack straw bales with gaps between them (like houses of cards) against the other two walls. Then, if the sow does move over there at any time, the piglets can escape into the straw. Lying on the piglets is no wilful act on the part of the sow; it is merely that she is so big and they are so small that she can crush them all too easily.

If you have a power source at or near the pig shed, then it can be a good idea to install an infrared heat lamp (with a red bulb) suspended in one corner, where the piglets can reach but the sow cannot. This provides the piglets with a warm corner to huddle and means that they are then only likely to bother the sow when they want feeding, since they do not need to seek warmth from her too. This is another way to help prevent them getting crushed by her lying on top of them.

The sow must be fed well throughout her pregnancy and while she is feeding her young, and the piglets should be weaned when they are four weeks old. Male piglets should also be castrated at this time, otherwise the pork will be tainted—a condition known as boar taint, which means that meat from intact adult males has an unpleasant odor released during cooking. Castration is a job best left for the vet.

Selling your pigs

Pigs can either be sold when they are about 12 weeks old, when they generally weigh about 90 lbs (40 kg) and are known as porkers, or kept until they are about 160 lbs (72 kg), when they are known as baconers.

You will soon discover that breeding pigs and fattening them are two different skills. It may be that you are good at both of them, but more usually you will find that you are better at one than the other and can decide to specialize.

If you keep your pigs to fatten them, you must take them to a local slaughterhouse to be killed and butchered. Some breeds will reach a weight of 260lb (118kg) and will yield about 70 percent of that weight in meat, which usually means you'll have more meat that you need. You can then keep any of the cuts of the animals that you want for your own consumption and freeze them, then sell the rest, either as cuts or processed into other goods.

BREEDS OF PIGS

There are hundreds of breeds of pigs. Some counties and regions have their own traditional local breeds of pig. Many of these are now considered rare or endangered breeds, but they are often hardier than the modern commercial breeds available. Keeping your own animals can be a good opportunity to revive this aspect of your local farming tradition.

HAMPSHIRE
One of the oldest breeds in America. Black-bodied, with a white belt around the front of the body, Hampshires have high-quality meat.

BERKSHIRE
Black pigs with short white snouts and white feet. Very high quality meat; good siring ability. Medium-sized animals.

DUROC
Durocs grow quickly and come in a range of colors from pale tan to dark red. Strong hogs with good meat for eating.

YORKSHIRE
Hefty pigs that came from Yorkshire county, England, and sometimes called Large Whites. Long-legged, large-boned animals.

CHESTER
White hog that is good for breeding, as the females make large litters and are good mothers. Medium-size; easy to handle.

LANDRACE
White hog originally from Denmark. Like the Chester, also a good mother. Long-bodied, with short legs; drooping ears.

Keeping Goats

Goats are far better and more practical backyard animals to keep for milk than cows. They are cheaper and easier to care for, more adaptable, require less space, and, on the whole, demand much less work and attention.

In return for your little time, goats give adequate milk yields for the average family. High-yielders will give up to 14 pints (8 liters) in summer, though this usually goes down to 2–3 pints (1.2–1.8 liters) in the winter. Providing they are properly fed and cared for, and milked in hygienic conditions, their milk will not be smelly or highly flavored, as it is sometimes said to be.

In fact, few people could tell goat milk from cow milk in a blind tasting, yet it has the advantage that it does not carry many diseases and is considered closer in composition to human breast milk. Goat milk makes excellent cheese and yogurt, which may be why more people around the world drink goat milk than cow milk.

The claim that goats smell is really only applicable to male goats (billies) and it is not advisable to keep these anyway. They can be very temperamental and are not productive for the homesteader—they are only useful to mate with female goats (nannies).

There are goat societies in both the United States and Canada. Contact a local branch to find out more about goat-keeping.

LIVING REQUIREMENTS

Goats need a dry, sheltered shed, where they can stand up, turn around, and lie down. They are determined escapees, so any fences around their enclosure must be extremely strong and well-erected to withstand their efforts to get out.

The size of your goats' shed will depend on how much room you have and how many goats you intend to keep, but a shed 10 x 15 ft (3 x 4.5 m) will give ample space

Goats naturally forage on bushes and other greenery, and like to reach up for their food, so always provide hay for them in hay racks mounted on the wall of their shed, rather than in mangers at ground level.

GOAT ACCOMMODATIONS

The way you choose to keep your goats will be governed by the amount of space you have. If you allow them to graze outdoors, they will strip an area of land of its plant growth, making them useful ground clearers and biological weed control. In time, you will need several large areas to rotate if you choose this option. Some goat owners will loan or even rent out their goats for weed-clearing, tethering them on a long rope to graze for the day or for several days.

SYSTEM	HOUSING	FEEDING	OTHER REQUIREMENTS
Outside	Dry, windproof shed. Deep bedding. Should equal outside temperature to encourage growth of winter coat	Roughage available from grazing. Provide hay at night. Give feed when giving milk	Salt licks. Access to water. Frequent inspection and handling to keep tame
Strip grazing, rotational grazing	Provide shelter although undercoat will protect from rain. Shut in at night	Allows for fresh grass feeding. Keep feed in dry containers and hay in well-ventilated shed	Electric fencing with three wires 15 in, 28 in, 40 in (40 cm, 70 cm, 100 cm) above ground. Salt licks
Tethering with shed (not for young goats)	Roomy shed for when not tethered. Platform for play and exercise	Move frequently to allow fresh grazing with bushes and branches available. Provide hay if there is a shortage of feed	Wide collar; tether-chain; swivel, harness, and stake. Shade in sunny weather. Access to water and salt
Yarding with shed	Both shed and yard as large as possible. Fence or wall so they can see out but not get out	Provide good variety, a garden for greens and grass from the edges	Take for walks for exercise and food. Salt licks and seaweed mineral supplements
Indoors	Roomy pen with light and air. Sleeping platform	Branches tied and suspended from ceiling. Scraps of bread and vegetables	Walks for exercise

Clipping goat hooves

Goats' feet must be trimmed every three weeks or so, or they will grow too long. The operation is best done by two people, with one to talk to and soothe the goat while the other does the cutting. Proper shears make the job easier—the horn is quite tough—although it can be done with a pruning knife and a steady hand. Always cut the hoof carefully.

Make sure that any fences around your goat pen are very strong and securely fixed into the ground so that the goats cannot knock them down and escape.

for three adult goats to live and get sufficient exercise.

Goats hate cold, windy, or wet weather and are best kept inside the shed on days like this, so a divided stable door, the bottom half of which should be about 4 ft (1.2 m) high, will offer some shelter, security, and ventilation at the same time. If there is a window, it should be positioned high up and preferably hinged at the bottom to give ventilation without being too drafty.

The floor should be concrete, and slope slightly towards the back of the shed to aid drainage (there should be a couple of drainage holes in the wall for this, too).

There should be a hay rack on the wall; hang it quite high, as goats are browsers rather than grazers, and therefore naturally reach upwards for their food. They do not like to eat off the floor. The floor itself should be kept covered in straw; keep adding to this as it gets dirty and then clean it out when the floor level has raised by 12–18 in (30–45 cm). Stack the dirty straw neatly and let it rot—it makes excellent garden compost.

Keep your goats in one place

Goats should be shut in their shed at night, but during the day they can be kept in one of two ways: either in a well-fenced run, or tethered in places that provide suitable food.

If you choose to keep your goats in a run, the area will have to be left empty occasionally to give it a chance to recover, so you must have at least a couple of such areas to devote to the goats. Fencing must be very strong: either erect solid, closely spaced board fencing or drive thick wooden posts 2 ft (60 cm) into the ground and secure strong wire fencing between them. Choose chain link, hog wire, or electric wire, or goats will merely trample their way through it, especially bucks. Make sure that the fence is at least 5 ft (1.5 m) high.

If you have access to lots of open land, where goats can safely be tethered, or alternatively have a lot of spare, scrubby land you want grazed, you can lead your goats out from their shed and tether them in a different spot each day. Goats will usually accept wearing a collar and being walked on a leash or length of rope.

Make sure there are no poisonous plants in reach where they are tethered, such as rhododendrons, privet, laurel, yew, milkweed, nightshade, mountain laurel, poison ivy, or poison oak. Common ivy is safe for them to eat, and they find oak leaves delicious. Goats will also make short work of flowers, vegetables, and fruit trees, so keep them well away from any of these that you value.

In winter, it is often advisable to keep goats in a paved yard, as muddy conditions can lead to foot troubles. Provide them with food and tether them out whenever the weather is good enough.

ACQUIRING GOATS

It is not fair to keep a single goat. They are sociable, gregarious, and friendly, and one on its own will be miserable. Over two, have as many as you want, according to the land and time you have available and the amount of milk you need.

You can buy young female dairy goats from any age—as young as four weeks, if you have lots of time to devote to looking after them and do not want milk for 18 months or so. If you want to be able to milk a goat immediately, you will need to choose one that has kidded—she will go on producing milk for about two years after this.

The advantage of getting a young goat, apart from the cheaper price, is that she will get thoroughly used to you, particularly if you give her the loving care and attention she likes to receive.

FEEDING

Another misconception about goats is that they will eat anything. In fact, it is when they are forced into a bad diet that they give bad-smelling and bad-tasting milk.

Goats are actually very fussy eaters and will not, for example, eat anything that has dropped on the floor of their shed. They require a meal each night and morning of some pelleted or textured complete goat feed. This should contain bran, crushed oats, and flaked maize, and possibly any of the following—peas, beans, molasses, barley, linseed cake, and salt. This feed should be mixed with some root vegetables and greens, including alfalfa or clover if you have any.

Use your vegetable scraps, although nothing that has been taken into your own kitchen for preparation. Make friends at a local grocery store and persuade them to let you have the offcuts from greens and any rotting dessert apples and moldy oranges. Midday snacks of beet are also appreciated. Always feed goats in a metal bucket or bowl—they can nibble at plastic.

How much to feed

The quantity of food to give a goat depends a lot on the age, size, and temperament of each individual goat and what else she is eating during the day. Find out what her daily diet has been when you buy her, and gradually change or increase this as necessary.

Feeds are important for milk production, but if you feed too much they will be wasted, because they

If you buy goats as kids, you will need to spend a lot of time caring for them in their early weeks, but they will thrive on your love and attention and will learn to trust you.

will not increase milk production over and above what is the normal average for that particular breed.

Clean drinking water must be readily available and the container will need frequent washing and cleaning. Goats will not drink from a slimy bucket.

Make sure hay is always provided in the hay rack in the shed, particularly if they are going out to eat on lush pastures during the day. Far from providing goodness, rich grass will just fill an empty belly with water and can cause bloating. It is unwise to let goats graze on really lush meadows for more than one hour a day. They also need a salt or mineral block to lick.

BREEDING AND MILKING

Although it is not uncommon for a maiden goat to come into milk, goats will not normally do so until they have kidded.

The first mating should generally be left until they are 15–18 months old. Some people claim goats can be mated as young as 7 months and, providing they are properly fed, will suffer no ill-effects thereafter; others say this will stunt their growth.

When they come into season—indicated by wagging of the tail, loud bleating, general nervousness and unease, and maybe a slight mucus discharge from the vulva—they should be taken to a billy as soon as possible.

Take advice from other local goat owners and make sure you choose a good billy to mate her with. The highest success rate in matings occur early in the nanny's season. They come into season every 21 days, usually from fall to late winter.

MILKING A GOAT

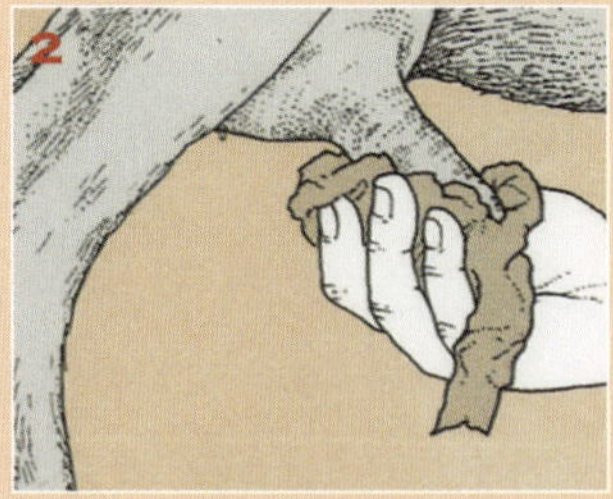

1 Before you take the goat to the milking shed, brush her coat to prevent any dirt falling into the milking bucket from her coat. Get her settled in the milking parlor and give her some food to distract her attention.

2 Wash the udders with a clean, sterilized cloth. Run your hands around the udders so that they are warm and the goat is used to your touch.

3 Collect the first milk from each teat in a small bowl or cup. This cleans and clears the flow. Discard this, as it will turn the rest of the milk sour very quickly.

4 Collect the milk you want to use in a clean, stainless-steel bowl: a traditional milking pail for a cow will probably be too tall. Squeeze the teats unless they are small, when they should be pulled up and down.

Gestation is about 5 months (155 days); if you are already milking the goat, stop two months before she is due to kid. Goats generally manage the birth quite happily if left alone, and usually produce two kids. They are usually taken away from their mothers soon after birth and hand-reared so that you can keep milking the goat yourself.

If one kid (or both) is a billy, you will need to decide whether to get rid of him, or rear and fatten him for the table—billies do not make suitable goats to keep long term and, obviously, are no use for milk production. The goat should be killed at about four months old and will taste a bit like lamb.

Milking goats

Milking must be done night and morning, every day of the year: the loud bleatings will tell you when it is time. You should have a milking shed, separate from the goats' living quarters, though not necessarily far away. It is important that this milking shed should be kept immaculately clean at all times.

Goat milk is very tasty, but you can also make delicious soft goat cheese by separating the curds from the milk and straining them through cheesecloth.

BREEDS OF GOATS

The breed you have is largely a matter of personal choice, although some are renowned for giving higher yields than others, such as Saanen and Alpine. It is best to be guided by what is available locally and by the advice of knowledgeable goat owners in the district.

Toggenburg
Large goat with long legs. Domesticated. Gives a high yield.

Saanen
Long legs, but baggy udder. Prefers good land. Quiet goat, which is easy to handle.

Nubian
Floppy-eared goat that gives good, low-fat milk yield. Flesh good for eating.

Alpine
Big black and white goat that gives a high milk yield. Good for grazing.

A good procedure for milking is to brush the goat to remove all dust, wisps of hay, and other scraps or debris from her coat that might otherwise fall into the milk. Then take her to the milking shed. You will find it easier if you have a raised platform on which she stands for milking (goats are very low to the ground). Use a stainless-steel bowl to collect the milk as you work—buckets are often too tall to fit under a goat.

The milking technique is to gently squeeze and release the teats—a bit like squeezing a sponge—rather than to pull them up and down. If your hands are big and the goat's teats are small, however, you will find you have to pull the teats up and down. As you get near the end of the supply, massage the udder and squeeze the teats hard to bring down the final drops. You will not hurt the goat, however hard you squeeze.

Using the milk

The milk should be covered immediately with a clean dishcloth, and then the goat taken back to her shed before she gets a chance to kick over the bowl. Strain the milk right away through a sterilized milk filter or pasteurize it in a home pasteurizer (available from livestock suppliers) and cool it by placing it in its container in a large bowl of very cold water or in a fridge. Then put it into a churn or jugs: it will keep fresh for two or three days in the fridge, but it must always be kept covered as it picks up other flavors very easily.

Alternatively you can freeze goat milk as soon as it is cold—it will not separate and will keep for up to six months—or you can make cheese (*see* page 236).

Keeping Chickens

Chickens are possibly the easiest of all animals to keep and must be about the most productive. Not only do they provide you with tasty fresh eggs, but once they have outlived their egg-laying usefulness, they make excellent eating themselves.

A small group of chickens need not take up much room or much of your time and they can be quite economical to feed. However, be prepared for a fairly high initial outlay. Proper chicken coops are extremely

LIVING REQUIREMENTS

Before your chickens arrive, you must decide how you are going to keep them and make sure that their new home is ready for them to settle in and start laying. There are many different ways of keeping chickens, and they all have their pros and cons; you just need to decide which is most convenient for you.

OPTION	HOUSING	PROS	CONS	FEEDING AND OTHER REQUIREMENTS
Free-range	Chickens are allowed to roam freely during the day, but are shut in a coop at night	This is undoubtedly the best way to ensure healthy, happy birds and will provide the best eggs. If chickens are allowed to scratch and peck in the garden, your food bills will be lower	The chickens will soon turn your garden into a meadow and will eat all the seedlings in any vegetable bed and many new shoots in the garden's borders, too. This is seldom a practical option for backyard farmers	Little additional feed required. Ensure a supply of clean, fresh water and make sure that shelter is available
Movable chicken coop	Chickens are kept in movable houses with a covered run attached. The house should be moved daily to give access to fresh ground	Less destructive than free-range and safe from predators	You will need a large area to accommodate the moving run. An orchard is ideal, but a rare luxury	Ensure a fresh daily supply of green food, either fresh grass or supplementary feed. Make sure that the run is on level ground that does not leave gaps at the base for predators to get in
Permanent chicken coop with run	A wooden house with space for roosting and laying is attached to a permanent, fenced run littered with straw	Space-efficient and secure. Can be constructed by customizing an old garden shed	The ground will become bare and unattractive and will need regular maintenance to prevent it from getting muddy in wet seasons	Feed the birds waste vegetables and scraps together with grass clippings or green shoots. Keep the feeder in the coop and provide a box filled with fine dirt or sand for scratching

expensive to buy, chicken wire for fencing is very costly, and the birds themselves are by no means cheap.

There are ways to economize, though: it is easy to customize an old shed with perches and places for the chickens to lay, and if you let your birds roam freely you will not need the expensive fencing.

One of the great advantages of keeping chickens is that it is very easy for someone to look after them for you if you want to go away. While people might be reluctant to come and milk goats or feed pigs, they are almost always willing to shut the hens in at night, let them out in the morning and feed them, in exchange for the eggs the hens lay while they are in charge.

A less pleasant aspect is that, even more so than with other livestock, chickens are prone to various diseases and ailments, so you must be prepared to wring a bird's neck when it is necessary. In addition, it is not uncommon to go to the coop in the morning and find that a chicken has died in the night for no apparent reason. Just remove it, but do not eat it, as it will not have been bled—the blood will not have drained from its veins as it does when you wring its neck and hang it (*see* pages 170–71).

Chicks have instant appeal, but are not always the best way to acquire chickens, as they require a lot of care until they start laying.

ACQUIRING CHICKENS

There are various ways of getting your chickens. You can buy them as day-old chicks, point-of-lay pullets, or older battery hens that have passed their peak of laying for a commercial egg-production operation but will still continue to lay eggs for some time.

Keep an eye on the classified ads and spread the word around other local homesteaders if you want to buy some chickens; there will often be someone nearby with chicks or pullets for sale.

Day-old chicks

These are the cheapest to buy, but they have to be fed for up to six months before they begin to lay. The mortality rate is often quite high and you do not know how disease-prone they are likely to be. They also need lots of care and attention. Some experts advise that they should be kept in some sort of incubator, although this is not necessary if you are prepared to keep them in a warm box in the kitchen, out of the way of cats and children.

Point-of-lay pullets

These are chickens that are about six months old and have just come into lay. They will lay for you as soon as they have settled down, which

BREEDS OF CHICKEN

Deciding which breed or breeds of chicken you choose is largely to do with whether you want colored eggs. The best advice is to buy locally whenever possible and choose a breed that is known to do well in your area.

RHODE ISLAND REDS
Fairly large birds able to stand tough winters. Prolific layers.

PLYMOUTH ROCK
Pretty, plump birds, good for eating. Lay light to medium-brown eggs.

ORPINGTON
Fairly large, with yellow feathers. Makes a good backyard bird.

LEGHORN
Smaller birds that are excellent layers of white eggs. Popular commercial type.

AURACANA
Medium to large chickens that lay beautiful blue-green eggs.

can be up to two or three weeks, but is usually only a matter of days. By and large, they are the best option, particularly for novice chicken-keepers. They are likely to be healthy and, providing you keep them properly, should remain that way. They are the most expensive to buy.

Choosing your chickens

Modern breeds of chicken have been developed with maximum egg production in mind for commercial use. Many of them will not go broody, so if you do want to breed from your chickens, it is better to choose one of the old-fashioned pure breeds.

Unless you want a lot of surplus eggs, you can estimate how many chickens you need on the basis of two hens per egg-eating person in the family. Be prepared for your egg consumption to go up if you keep chickens, as your supplies will be replenished daily.

The only real reason for keeping a rooster with hens is if you want to produce chicks, although it is a good idea with free-range hens, as a cock will keep his hens in order, bossing them around and apparently pecking them at random. A rooster has a far more voracious appetite than his hens, so is much more expensive to keep, and if the hens will go broody, it is possible he will turn the whole pack so at one time. In addition, his early-morning crowing can quite often upset neighbors.

FEEDING AND DAILY ROUTINE

As with all livestock, keeping chickens has to become part of your own daily routine.

If the chickens have a run, open up the coop and let the hens out as

A chicken feeder releases a little grain at a time, as the supply in the bottom tray is eaten. Alternatively, scatter feed over the ground and let the chickens peck for it.

early as possible—dawn is best. If left too long, the birds may start to peck each other. Feed them and give them clean water.

Collect the eggs in the morning (most are laid between nine and noon and should be collected as soon as possible after laying to avoid breakages). Unless you are cleaning out the coop or digging the run, the chickens need no further attention until their next feed at three or four o'clock. Shut them into the hen house in early evening to keep them safe overnight.

Feeding requirements

What you feed your chickens depends on how you are keeping them and what they can forage for themselves. Those kept in a run will need some chicken feed night and morning. This is easiest (and most expensively) supplied in the form of commercial layers' mash or pellets. Pellets are the most economical, because mash, even when wet, tends to blow away in a strong breeze. Instead, you can feed them about 4 oz (125 g) of scratch (mixed corn or wheat) divided between the night and morning, and supplement this with a good supply of household scraps. To avoid contamination and disease, only give vegetable scraps, never meat, and only use vegetable trimmings and scraps that you have prepared outside your household kitchen.

Concentrated feed can be scattered on the ground for the birds to scratch, or fed in a container. Clean water is essential and chickens drink an enormous amount. You can buy special water containers, which are designed to help keep the water clean. A cheaper alternative can be made from a plastic bowl with an upturned terracotta pot placed in it. This should stop the hens from getting into the bowl and making it dirty.

Grit and limestone

In addition, chickens need grit, which helps them to break up their

Eggs

Hens lay most eggs in the summer when the days are longer and lighter. In their prime, you can generally expect to get eggs for five or six days a week from one chicken. This will drop off in winter, although from a flock of about 10 birds, it is usual to get one or two fresh eggs each week, even in midwinter. Hens can be encouraged to lay in the winter if you light the coop well, so they have, in effect, about 12 hours of daylight each day.

food, and calcium or limestone to help them form egg shells. This can be crushed oyster shells (available from a local farm supply company) or recycled egg shells, dried off in a low oven and crushed. Each bird should have approximately ½ oz (14 g) of grit and limestone (or oyster shells) per week in the ratio of one to four grit to limestone.

BREEDING

If you have a broody hen and a rooster, you might like to try to hatch some eggs. It is fun to do, lovely to watch, and you may be able to add to your flock.

Broody hens are instantly recognizable—they refuse to move from the nesting boxes, sitting tight and clucking gently, they stop eating food, and they lay no eggs. Warm weather from spring onwards is the time to watch for this. If you do not want a hen to be broody, put her in a cold box on her own, with no straw so she cannot make a nest, and place her where she can see the others. Let her out at feeding time and check a little later whether she has gone back to the laying box. If she has, put her back in the isolation box for a while longer.

When a hen goes broody and you want to breed some chicks, put her in a special box of her own, this time away from the others and with some straw so she can make a nest. She should not be able to get out of this, and ideally it should have a wire mesh bottom to allow moisture and warmth to come up from the earth. Sit her on eight newly-laid eggs and leave her. Take her off the nest to feed and perhaps have a dust bath, but the eggs should not be left for more than about 15 minutes or they will get too cold. Providing all goes well, the chicks will hatch exactly 21 days after the hen began sitting on them. There is nothing for you to do but wait until the tiny yellow chicks emerge.

Chicks will hatch three weeks after a broody hen starts sitting on the eggs. Introduce them to the rest of the flock gradually until they are accepted, so that they do not get pecked.

Looking after chicks

The chicks need water but no feed for 24 hours. Put the water in a tiny container, otherwise they will fall in and drown. Thereafter they can be fed on chick pellets three times a day, or you can feed the mother grower's mash and watch her feed tiny crumbs to her babies.

You will be able to tell when the chicks are ready to leave the hen and become independent, because she will let you know she is clearly fed up with them. It is best to integrate them into the flock slowly, if you can, as they are likely to get pecked and bullied at first. Cock birds can be fattened for the table.

Hens are not always good mothers. If you get a good one, keep her for as long as possible to rear chickens for you. It is best, though, to change the rooster each year. If you keep the same one, your flock will begin to be inbred.

Clipping wings

Chickens should have the flight feathers on one of their wings clipped when they are six months old, or they will fly over fences and escape, perhaps into danger. Spread out one wing: you can clearly feel the meaty part of the wing and will be able to see the long flight feathers. Clip these off at the base, using very strong scissors, as the feathers are very tough. Although it seems a brutal operation when you do it, it will not hurt the bird or draw blood.

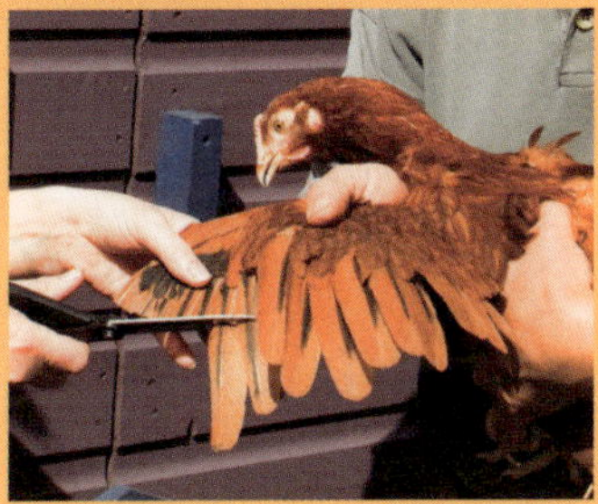

PLUCKING AND PREPARING A CHICKEN

There are two processes in preparing a dead bird either for the table or to be frozen before eating: first, it must be plucked to remove the feathers and then it must be "drawn" to remove the innards.

Plucking is easiest done while the bird is still warm, as the feathers will come out more readily than when the carcass is cold. However, domestic hens and other poultry will have been toughened by a lifetime of pecking and scratching, and some people feel that the taste and tenderness of the meat is improved if it is hung for 24–48 hours first.

Hanging poultry

Hanging is usually done before the bird is plucked or drawn; if the feathers are left on, fewer flies are attracted to the carcass and the bird tends to keep better with the innards intact. You can hang poultry after plucking and before drawing if you do so in a cool, airy place where there are no flies.

If you want to hang the poultry for a day or so, you can make the subsequent plucking an easier job by pouring a pot of boiling water over the bird—just to warm it very slightly on the surface and loosen the hold of the feathers a little.

Plucking a bird

Plucking is a messy job, so lay down plenty of newspaper or tarps to catch any flying feathers. When you are ready to pluck your bird, put it on a clean work surface and pull out the long tail and wing feathers first, jerking them sharply. Pluck systematically from the neck, back, breast, wings, and legs, taking a number of feathers between your thumb and fingers each time and tugging them sharply. The tiny pin feathers that remain are easiest to remove using tweezers or by grasping them between your thumb and the edge of a blunt knife.

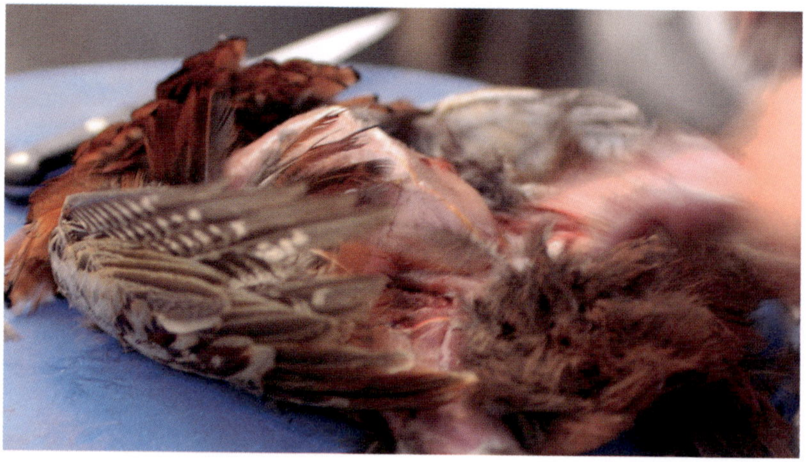

If you pluck a bird promptly after it has been killed, the job will be quicker and easier, as the feathers will come out more readily. Work methodically to do a good job.

Down feathers

All poultry is plucked in the same way, but ducks and geese are harder and take longer, as they have a covering of down beneath their main feathers. This down is removed in the same way, but is fluffy and will fly around as you work, settling in your hair and on your clothes. You can scald the feathers first by pouring over some boiling water, and this will reduce the problem.

Collect the down, wash and dry it (in paper bags in a low-temperature oven or by hanging in a warm and airy place), and when you have enough, use it for stuffing pillows.

Plucking a goose is hard on your hands; a good tip is to wrap a bandaid around the forefinger of your plucking hand to cushion it.

Cutting up a chicken

With the breast side down, insert a knife at the bottom of the neck and slit the skin up to the head. Pull the skin away from the neck, then cut through the neck bone as close to the shoulder as you can, keeping the knife inside the skin.

Remove the crop and the windpipe (trachea), which make up the giblets, pull out the neck bone, and cut around the skin close to the head to separate it from the body. This way you leave a large flap of skin that can be folded over the bird's back and will help to prevent the meat from drying out as you roast it. Discard the neck bone and head.

Turn the bird around, insert the knife into the vent and make a hole. Loosen the innards by putting a couple of fingers into this opening and then into the opening at the neck end and gently working them round against the inside of the carcass. This job will be less unpleasant if you have starved the bird for 24 hours before killing it.

When you feel the innards are loose, gently pull them out through the opening at the vent. If you keep them together and are careful, you should not break the gallbladder. If it breaks, it gives the bird a bitter taste.

Run cold water through the bird to clean it inside and out and dab it dry with paper towels. Remove the feet by slitting the skin round the knee joints and twisting the lower legs. Cut through the white sinews. Tuck the flap of skin at the neck under the wing joints.

Separate the heart and liver from the rest of the innards, taking care not to damage the gallbladder attached to the liver. Cut open the gizzard and wash it. This, with the heart and neck, forms the giblets and can be used to make stock or soup. Use the liver separately.

Killing birds

The most economical way to keep chickens in terms of getting the maximum egg production is to keep them for two laying seasons (until they are about 18 months old) and then kill them and start the flock again. They will go on laying for some years to come, but although the eggs get larger, they will also become fewer.

The best way to kill a bird is by wringing its neck. Stun it with a blow to the head then hold the legs with your left hand, letting the body hang down. Hold the neck close to the head with your right hand. Pull the neck down and twist firmly so that the head is bent backwards, towards the back. You will feel the neck break. If you pull too hard you will pull the head off. Hang the bird with its head down so the blood drains out of its veins before plucking and preparing (opposite).

Ducks, Geese, and Turkeys

If anything, ducks are even easier to keep than hens, and some people say they are friendlier. Geese can be aggressive but they are good foragers. Turkeys are relatively self-sustaining.

DUCKS

Although some people say that ducks will be content with just enough water to immerse their heads and necks, you really need a pond, river, or lake to keep these birds happy.

Ducks' living requirements are far more primitive and straightforward than chickens. All that is necessary in addition to a body of water is a dry house containing some straw, in which they can be shut up at night to protect them from predators. It needs no nesting box or roosting perch—in fact, a wooden packing case with a couple of ventilation holes drilled in the back can be turned into an excellent duck house.

Keeping ducks for eggs or meat

Duck eggs are often considered to contain impurities, which is one of the reasons why ducks are less frequently kept than other poultry. In fact, the shells are porous, so if laid in the mud around a stagnant pond, dirty water could get into the egg. Laid in clean surroundings, they are perfectly good, clean, and quite safe to eat. The best way to ensure this is to keep the ducks shut in their house until about mid-morning. They lay their eggs before this time and therefore will do so in the clean house and not in the mud.

The eggs are larger and richer than hens' eggs, but will not keep as long and should be used within a week.

If keeping birds for the table, buy them when they are very young and keep them for 10 weeks. Fatten them with barley meal and kill them when they are exactly 10 weeks old, at which point they are at their best.

Don't attempt to keep ducks unless you have a large enough pond or lake for them to swim in, or have a stream running through or beside your property.

They are also easiest to pluck at this stage; thereafter, they begin to molt and the new feathers, which will be starting to come through, are much more difficult to pull out.

If keeping them for eggs you can get ducks as day-old chicks or point-of-lay pullets, just like chickens (*see* page 167).

Feeding

Ducks are voracious eaters, but will thrive on little more than plain grass and will help you in the garden by keeping down slugs and snails. They will also keep down the vegetable garden by eating your crops, so make sure they cannot get to it. They will need some grain in the night and morning. Feed it in a container rather than scattering it and if there is any left after an hour, feed them less next time. They will also appreciate their share of greens and vegetable scraps. Ducks need as much clean drinking water, grit, and limestone for their shells as you would give to chickens.

Breeding

In general, ducks do not make good mothers. It is much better to give a clutch of eggs to a broody chicken and let her hatch them. This will take 28 days, but dampen the eggs each day in the final week, as ducks would do.

Do not let the chicks into the water to paddle or swim until they are four weeks old—before this, their feathers will not repel the water and they could easily drown.

If your ducks are an egg-laying breed and you have extra drakes, kill them at 10 weeks old and eat them. Ducks have their necks wrung in the same way as hens (*see* page 171).

A duck house does not need much enhancement—a box will be quite adequate—but a ramp sloping down to the pond will be appreciated, both as a slide and an exit.

GEESE

If you have a large orchard or patch of grass where they can graze, or if you have access to unlimited greens, geese are probably the cheapest of all poultry to keep.

Although they have become less popular, geese still make excellent eating and many people prefer them to turkeys at Christmas. There are a number of different breeds, most of which are bred for meat rather than eggs. Common breeds include Sebastopol, African, Chinese, Pilgrim, and Toulouse. Goose eggs are larger and richer still than duck eggs, and you can expect to get about 100 eggs a year from each of your birds.

An important point to consider before buying some geese is that they are extremely noisy and aggressive and will chase after people they do not know. This can be both an advantage and a disadvantage, but they are well known as excellent watchdogs against intruders.

If you keep geese, you must establish from the start that you are the boss, particularly with the gander. Look him in the face and let him know you are not passive and unsure. The way to deal with a goose if it does attack you is to grab it by its neck. This is easier than it sounds, as long as you are bold, as the bird advances with its neck forward, squawking at you as it approaches. It is unable to do anything if you grab it firmly, but not tightly enough that you will hurt it, around the neck.

Living requirements

Like ducks, geese have very simple housing requirements and only need somewhere with four walls, a roof, and a floor, in which they can be shut up at night to keep them safe from predators. Alternatively they can be kept in a well-wired run (put wire on the top as well as the sides, for extra protection).

Feeding

Their staple diet is grass, but they cannot cope if it is really long; it needs to be of medium length, which they will crop as neatly as a lawn. If you want to fatten them, give them some grain too, and if you put this in their overnight accommodation, it will help to lure them inside. In addition, like all poultry, they need grit and limestone, and access to clean water. They drink great quantities of water, in fact, and as they immerse their entire heads as they drink, it should be given in a fairly deep container. Clean it as soon as it gets dirty.

Breeding

You can breed from your geese if you want to. They sometimes take a while to select a mate, but once done, they have paired for life. You can expect the geese to produce 10–20 eggs, but these will seldom all hatch in a single clutch. Beware of the male, particularly during breeding time; he usually gets very protective and aggressive and it is best to try to stay well away from the mother goose's nest.

Geese are not good mothers, so if you want to breed goslings, put 4–6 eggs at a time under a large broody hen. They take up to four weeks to hatch, and if being hatched by a hen, the eggs must be turned each day in order to stop the contents settling. Sprinkle them with water, making sure they are kept wet in the final week before hatching. Feed goslings on bread and milk to begin with and then on chick feed.

Geese will peck at grass on the ground, but can also be fed supplementary grain if necessary. They are noisy birds, so do not keep them too close to your neighbors.

Killing geese

Geese cannot be killed in the same way as chickens and ducks: they are too big and strong. There are two ways of killing them, but for both, you must stun the bird first by giving it a hefty blow to the head with a heavy instrument. Then, either cut its throat at the bottom of the neck to sever the jugular vein, or hold the bird by the feet, with its chin resting on the ground. Get a helper to lay a pole or metal bar across its neck and tread on either end of this to keep it in place. Then pull the legs up towards you to break the neck.

TURKEYS

These are the birds least frequently kept on a small scale, mostly because they have a reputation for being hard to raise. They are usually bred solely for their meat, although their eggs are actually very similar in taste to those of chickens. One turkey egg is approximately the equivalent in size to two hen's eggs.

BREEDS OF DUCKS AND GEESE

There are some traditional local breeds, which make a good choice if you just want a handful of birds. Contact local suppliers to find out which breeds do best in your area and which are most widely available.

EMBDEN
Large pure white bird—the most popular breed of goose.

CHINESE
White or brown goose, which is smaller than the Embden.

TOULOUSE
Slightly smaller than Embden, with mostly gray feathers.

AFRICAN
Goose with relatively lean meat. Brown or white varieties, both with calmer temperament.

AYLESBURY
Very popular duck, which is kept for eating. White feathers.

MUSCOVY
Good meat duck that does not quack, but hisses.

Turkeys are prone to get a disease called blackhead if they come into contact with chickens, so they cannot be kept together in the same run. In fact, this contact can even be as remote as you dealing with the turkeys after having been in the chicken coop. However, modern turkey feeds contain antibiotics which combats this disease, so providing you give medicated turkey feed, you should encounter no problem. This feed should be withdrawn before slaughter.

A turkey house is similar to a chicken coop in design, although the birds are much larger. The turkeys like to have perches, but make sure these are strong and solid enough to bear their weight. It's better to keep the turkeys free-range. They will wander around all day constantly pecking and eating.

Breeding turkeys

Turkey eggs can be difficult to hatch; they are large birds for the size of eggs they lay and they are apt to break them inadvertently in the nest. It is better to use a broody hen if possible, and she can hatch from 6–8 eggs. She will then rear the chicks, teaching them to eat.

Using an incubator

An alternative is to hatch turkey eggs in an incubator, kept at about 101°F (38°C). Keep the incubator where the temperature is constant—in the house is better than in a shed, where the temperature will fluctuate. The trouble with hatching turkeys in this way is that young chicks are very difficult to rear as they are reluctant to begin eating without a mother bird to teach them. To overcome this, put chopped onion tops on their food.

Fattening chicks for Christmas

Besides there being a market for turkey eggs, the chicks can easily be sold as day-old chicks, or you can keep them, fatten them, and sell them at Christmas.

Turkeys are best kept as free-range birds, if possible. They will still need to be shut in a shed at night, though, to keep them safe from predators.

Keeping Bees

Bees cannot truly be "kept," but if you can persuade a colony to take up residence on your property, it is well worth doing so. Even one hive will provide you with all the honey you can eat.

Take advice from beekeepers in your area if you are a complete novice. It is a good idea to join a local beekeeping society; most have courses for beginners.

Can you keep bees?

Before deciding definitely to keep bees, it is important to discover how you react to being stung. Some people react very badly and keeping bees could be too dangerous. Consult a physician for allergy testing before subjecting yourself to the inevitable stings.

EQUIPMENT

Certain equipment is essential, and it can be expensive. Try to borrow or buy things secondhand when you are starting out—you can always upgrade later.

The most essential piece of equipment for yourself is a bee-veil to protect your face from stings. You will also need gloves (you may dispense with these later, but wearing them helps to give confidence to a beginner) and rubber boots, into which you tuck your pants, which are, themselves, tucked into long socks, so the bees cannot get inside.

You might prefer a bee suit, which offers complete head-to-toe protection, but these are expensive unless improvised out of a pair of overalls. Remember that bees climb upwards, and so will climb up sleeves but not down pants, and protect yourself accordingly.

Bee equipment suppliers sell beginner's kits. These include a hive, a hive tool to lever out the frames, and a smoker. This is used whenever you want to open up the hive; in it you burn anything that smolders, such as dried grass, pine cones, or half-rotten wood. The smoke subdues the bees, as their reaction to it is to gorge themselves with honey, making them dozy and therefore less likely to sting when you disturb them.

A complete bee suit (below) offers head-to-toe protection against stings. Using a smoker (above) will subdue the bees before you open up the hive to work inside, making them less likely to panic and sting you.

To start a colony of your own, you need a nucleus of bees, including a queen bee that is laying. Introduce this to your hive and the colony will build up to around 40,000 bees.

THE HIVE

This is the most expensive item of equipment, but may often be acquired secondhand or in a starter kit. Scrub and disinfect it before use, and check whether it needs any repairs.

Most beekeepers use a hive that has a frame designed for the bees to make their honeycombs. A model with easy-to-remove frames called the Langstroth is the most popular.

Positioning hives

Do not site hives near a sidewalk, walking trail, driveway, or road, where people could be bothered by the flight of the bees, nor under heavy trees where the spot is likely to be damp, or a frost pocket. Ideally, they need a place that is screened off from the rest of the garden.

ACQUIRING BEES

It is always best to acquire bees from a local beekeeper: the bees are adapted to a particular region and its local conditions. You can, however, also purchase bees by mail order.

The easiest way to start a new colony is to acquire a nucleus from another beekeeper. This is a mini-colony, consisting of about five frames of drawn comb with 5,000–8,000 bees and a laying queen. A beekeeper should not let you have this nucleus until it is certain that the queen is laying—early summer is most likely.

Getting started

The nucleus is put into the center of the lowest box—the brood chamber or brood box—which is filled on either side with frames of foundation. Another box of foundation is placed above this. The initial aim is to build up the colony, but the queen will lay according to the food coming in and as there will not be many spare bees to go out to collect nectar, you will need to give supplementary feed. Make a syrup from 2 lbs (900 g) of sugar dissolved in 1 pint (550 ml) of hot water, cooled to room temperature. The amount of feeding will depend on the weather.

It is unlikely that you will get any honey in the first year, but if the weather is right in summer (hot, sunny, sultry days with rain at night to encourage plant growth) you can replace the feeder with a super (a box of eleven frames) full of foundation frames. At the end of the season, you could get between 10–20 lbs (4.5–9 kg) of honey.

As the colony builds up over the first few weeks, move the frames around every two weeks or so, being careful not to split the brood. By winter you will have two full boxes of drawn comb which is ideal for over-wintering the bees inside.

Dealing with stings

If you keep bees, you will get stung sooner or later. Scratch out the stinger right away, using a hive tool or fingernail. The bee leaves a poison sac with the stinger, plus the mechanism that continues to pump the poison from the sac, so if you squeeze the stinger, it will only send the poison shooting into you, causing more irritation and swelling. Generally, the pain stops in a few minutes and the swelling subsides in a few hours. If it is really painful, bathe the sting with witch hazel or soak with diluted Epsom salts.

Managing the colony

Inspect the hive every few weeks from early- to midsummer to make sure the queen is laying and to see if there are any queen cells. These are much larger than ordinary cells and will protrude out from the comb. If you allow a new queen to hatch out, the colony will swarm.

Either remove the frame and destroy the queen cells, or make an artificial swarm (a natural swarm is when all the bees leave the hive and alight on nearby trees and structures). Find the old queen in the hive and put her and the frame of eggs she is on in another box with some frames of drawn comb. Leave these on the current site and put all the brood frames, including the queen cells together with the drones, on another site. When the foraging bees return, they will automatically

go into the old hive; this will house the queen and all the foraging bees, and is known as an artificial swarm.

Meanwhile, in the adjacent site, with the nurse bees and the queen cells, panic will set in as there is no food coming into the colony and the bees will tear down all the queen cells except for one. The new queen will emerge, leave the hive, get mated, and come back to start laying, beginning a new colony. If you only want one colony, find the original queen and destroy her or give her away. Put the new queen into the old hive and the bees will follow her. Renew the queen in a colony every two years.

REMOVING HONEY

Honey may be taken from the hive from midsummer to late summer, depending on the available nectar.

Put a board with a one-way bee escape under the supers and, when the bees have gone down to the lower boxes (after about 48 hours), remove the top frames. To extract the honey you must first scrape the wax cappings on the cells off with a large uncapping knife. The frames are then put into a special centrifuge or honey extractor. It spins the frames round at speed so the honey falls into the bottom of the drum. From there it is filtered into the honey tank, from where it can be drawn off into jars. Another method is to simply cut the honeycomb from the frame, then to cut smaller pieces from the comb. Separate the pieces and let the honey drain off, then place them in containers and seal.

Extract honey from all the boxes above the queen excluder. This leaves some honey in the lower boxes, but not enough to feed the bees through the winter, so give them a syrup feed from fall.

SWARMING

If you inspect your bees regularly and follow the procedure outlined above, experienced beekeepers would claim that there is no reason for bees swarming.

Bees will only swarm if the hive becomes overcrowded or if a new queen hatches. If the bees do swarm, about half of them will leave the hive with the queen. They will cluster nearby and, unless they are collected by a beekeeper, move off a day or two later. If you lose sight of the bees, another beekeeper may collect and claim them.

Your honey will have some of the flavor of the flowers the bees fed on, although in most small farms and gardens, they will have flitted from plant to plant, giving no particular taste.

Collecting a swarm

Place a white sheet on the ground beneath the swarm. Hold a strong cardboard box or straw basket directly under the largest part of the swarm and shake the branch or bush where the bees are resting. This will cause most of the bees to fall into the box. Invert the box over the sheet and prop up one side with a small stone. After an hour or so, all the bees should be in the box.

Return at dusk and wrap the box in the sheet, then take the swarm to a new hive. Place a board in front of the hive, sloping up to the entrance. Spread the sheet over the board and shake the bees out of the box. Bees run uphill when they are scared, so they will run up into the hive.

FOOD FROM NATURE

Gathering from the Wild
Herbs and Plants
Fruit and Nuts
Mushrooms
Fishing
Food from the Oceans
Hunting
Gathering Firewood

Gathering from the Wild

The fields, woods, inland rivers and lakes, and the shoreline with its coastal waters all harbor rich crops of edible plants, animals, and fish, and yet few of these are exploited to their full potential.

Always treat any produce foraged from the wild with respect and caution, but this does not mean it needs to be completely ignored. The country is home to all kinds of mushrooms, fruit, nuts, herbs, plants, and flowers, many of which will augment and enliven your dinner table and cost nothing more than the enjoyable time that you take to find and pick them.

Mushrooms may be the most neglected of all free food: millions of edible fungi go uncollected each year in forests, but amateur foragers are understandably reluctant to try them. Some species are deadly poisonous and unless you can positively identify them, there is no foolproof test that will tell you if a particular mushroom is safe.

It is worth learning to identify the poisonous ones, so that you can gather and eat the others. This chapter will help you, or you could join one of the many mycological societies organized around the country, where an expert will show you the ones that are safe—and tastiest—to eat.

Harvest from fields

There are many edible wild plants and herbs that grow in abundance. Be on the lookout as you explore parks and open land in your area, and ask people who have lived in the area for a long time whether they know of anything that grows in any nearby woods or meadows that is especially good to eat.

Many of the plants that abound along the edges of our roads today were attributed with great healing powers by indigenous peoples, and can still make delicious additions to soups and salads. But when harvesting wild plants, follow a few simple rules in order to ensure your own safety, as well as the survival of the wild and native plants.

Make sure that you know that what you are picking to eat is edible and not poisonous; if you cannot find someone who is definitely able to confirm this for you, leave the plant alone. Most produce—fruit, flowers, and fungi in particular—is best gathered on a sunny rather than a rainy day, as the wetness is likely to encourage it to deteriorate faster.

Place your harvest into an open shallow basket, rather than a plastic bag (in which they will be crushed), and pick only the quantities you know you can deal with at any one

Only gather edible produce from places that are free from smog, runoff, and other pollution. Pick only what you can identify.

Fishing might not be a practical everyday way to obtain food, but a day fishing off the beach or for freshwater fish can be fun, rewarding, and productive, too.

time. Try to gather only from the places that are relatively unpolluted. Most plants are best if gathered when they are young.

Proper respect for natural places and the wildlife there is of paramount importance. It is illegal to pull up wild plants completely by the roots in some places and, in most cases, it is unnecessary. Likewise, do not totally strip plants of their leaves and flowers—this could easily kill them. Nor should you take all the flowers or seeds from annual plants or they have no hope of being able to reseed and grow again.

Meat and fish

The outdoors is home, too, for a number of wild animals and birds that can be caught and eaten. There are, of course, extensive rules and regulations that govern all aspects of hunting, shooting, fishing, and trapping, so make sure you are aware of what laws you must comply with and obtain all the necessary permits or licenses before you start.

Rivers, inland lakes, ponds, and reservoirs can be fished to provide food, although this is often more an activity that is enjoyed as a sport than as a way of obtaining food. The equipment used can be sophisticated and expensive—far from the stick-and-bent-pin image of one-time riverside fishing.

The seashore and its immediate coastal waters is another rich source of edible goods. Around the shoreline you can find a number of plants, as well as various edible seaweeds and the many sea creatures that bury themselves in the sand or live among the rocks. Or you can try your luck with ocean fishing, crabbing, or lobster trapping.

Herbs and Plants

The list given below of edible roots, leaves, and flowers is by no means complete but illustrates some widely available and commonly used plants. Each part of the country will have its own local native flora that can be harvested.

Arugula, mustard, carrots, clover, dandelions, sunflowers, radish, salsify, violets, and watercress may all be found growing outside of cultivation, although probably not all in the same area. You may also find wild garlic, ramps, wild ginger, and edible ferns growing abundantly in woods in spring.

Beech (*Fagus sp.*)
Pick leaves when they are very young and tender. They make a lovely addition to salads.

Broom (*Cytisus scoparius*)
Considered a weed in many areas. Pick the flowers when they are still in bud and pickle them in vinegar or salt. Alternatively, sprinkle them over salads as a decorative and tasty garnish.

Burdock (*Articum sp.*)
Grows in abundance in thickets, roadsides, and in disturbed sites. Eat the young leaves as a salad vegetable; chop the leaf stems raw into salads, or simmer them gently and serve with butter as a vegetable. Scrape and boil the root to eat like salsify.

Chickweed (*Stellaria media*)
Grows as a weed in most gardens as well as on vacant lots everywhere. Strip off the tiny leaves by running

An abundance of chickweed makes a vibrant green carpet over this woodland floor—a marvelous haul for a forager.

a fork down the stems and cook them like spinach, or use raw in sandwiches or as a garnish.

Chicory (*Cichorium intybus*)
Grows in alkaline soil. The leaves can be used raw in salads and the root can be ground to make a drink that is often used as a substitute for coffee. To do this, clean the roots thoroughly and bake them in the oven at a low temperature until they are dry and crisp, then grind them in a coffee grinder. Let them steep for some minutes in boiling water before straining and drinking the liquid.

Comfrey (*Symphytum officinale*)
A much-neglected plant that grows commonly in ditches or damp ground near rivers. It makes a pleasant alternative to spinach and can be cooked in the same way.

Dandelion (*Taraxacum officinalis*)
One of the most common, yet most useful of all wild plants. Found growing abundantly in all wasteland and open grassy spaces (and many suburban lawns), it is generally considered to be a weed, but it is not so long ago that the plant was cultivated in vegetable gardens. The leaves make a tasty and nutritious salad green, but should be picked young—in springtime before the well-known yellow flowers appear—as this is when they are at their least bitter and most tender. They can also be cooked like spinach, and many people mix them with the leaves of spinach or sorrel (*see* page 184), both raw and cooked, in order to reduce the rather bitter taste. The roots can be dried in the sun, then roasted and ground to make a coffee substitute. They can also be cooked and used to make wine or cordial.

Lamb's quarters (*Chenopodium sp.*)
One of the common garden weeds, it will also be found growing on

Dandelion grows like a weed in many garden lawns and beds, but the leaves, in particular, are good to eat when they are young.

wasteland. It is rich in iron and is another substitute for spinach.

Garlic mustard (*Alliaria petiolata*)
An invasive herb that grows in abundance. The leaves can be chopped over salads (they have a slightly garlicky flavor), or made into a sauce to accompany meats.

Glasswort (*Salicornia sp.*)
Found on salt marshes, beach dunes, and salt flats in coastal areas. The young stem leaves can be cooked like any leafy green vegetable. Also called samphire.

Hawthorn (*Crataegus sp.*)
Many different species, of which the English and European types are used as garden trees and hedging plants. Flowers are traditionally used to make a fresh-tasting liquor by combining the petals with brandy and superfine sugar; berries can be made into jelly.

Horseradish
Vigorous garden escapee that grows in abundance on waste ground. Peel and grate the roots to make horseradish sauce or cream.

Lime (*Tilia sp.*)
The healing properties of this tree were such that many years ago they were planted by royal command along roadsides in the United Kingdom.

The leaves make a good salad vegetable or sandwich filling. The flowers, picked while they are in full bloom and dried, make a refreshing tea that is renowned for being extremely soothing.

Bright red hawthorn berries make a good jelly to accompany roast, red meats, or pick the flowers in spring to flavor brandy liqueur.

Miner's lettuce (*Montia perfoliata*)
Grows widely in moist, shady woodland and fields. All parts of the plant are edible. Put the leaves in salads or boil or steam like spinach. Was widely eaten by early settlers in the West.

Plantain (*Plantago sp.*)
Widely found growing on waste land, on disturbed soil, foothills, and in parks and gardens. The young leaves can be used raw in salads or cooked like spinach.

Poppy (*Papaver somniferum*)
Another plant whose contribution to the family pantry cannot be claimed as a great money-saver. However, the seeds contained in the brown seed heads that follow the red, papery flowers, can be used to flavor cakes, cookies, and bread, or sprinkled over salads. There may be restrictions on growing this poppy in some areas.

Primrose (*Primula sp.*)
The flowers of this spring plant can be used to make a refreshing drink or they may be crystallized in the same way as rose petals to make edible cake decorations. Gather them sparingly though: primroses were over-picked in the past and their numbers are only just recovering.

Purslane (*Portulaca oleracea*)
Also known as pigweed or little hogweed, this succulent is considered a garden weed. The leaves, stems, and flower buds are all edible. It can be put into salads, stir-fried, or cooked like spinach.

Salad burnet (*Sanguisorba minor*)
Commonly found in grassland, particularly in alkaline and limestone soil. Crush the leaves slightly and use them in salads, or add them to cool long summer cocktails or iced tea for a refreshing drink.

Sorrel (*Rumex sp.)*
Throughout the spring, wood sorrel may be found in woodland and shady places; common sorrel is found on grassland and open country. Use raw in salads, or cook in soups, soufflés, omelettes, stuffings, and sauces for meat and fish, or as a vegetable accompaniment. It contains oxalic acid, so do not eat it too frequently.

Stinging nettle (*Urtica dioica)*
Another much-neglected plant that is widespread and grows in great abundance. Pick it when very young—ideally no more than 6–8 in (15–20 cm) high—as you are less likely to get stung by young stalks, and cook them like spinach (the stinging formic acid is destroyed in cooking), use them as a flavoring for omelettes, or make them into soup.

Wild rose (*Rosa sp.)*
The flowers have a pleasant, if not very substantial, culinary use. They can be used to flavor jams, jellies, or honey. Painted with egg white and sprinkled with sugar, the crystallized petals make a pretty, edible decoration. Try them as an unusual sandwich filling or as a delicate garnish to a light salad. A traditional use for rose petals is to make a light syrup. For rose hips, *see* page 187.

Sorrel has myriad uses in the kitchen, in soups, salads, stuffings, and sauces. Look for it in spring, when it is most abundant.

HERBS

Only the herbs most easily found have been listed below. If you are lucky you will find lemon balm, borage, parsley, fennel, basil, and rosemary, and quite possibly others too. If you find a large patch of borage, you can use it as a salad vegetable, cook it like any leafy green, or as a base for soups. If you have a large enough quantity of herbs, you can preserve them (*see* page 219) by drying or freezing.

Angelica (*Angelica sp.)*
May be found in damp places around the outskirts of woods or near streams. Although classified as an herb, its main culinary use is as a decoration for desserts or flavoring for cakes. For this, the stems are candied by simmering and steeping them in heavy syrup over a period of a few days. Be sure you identify this plant correctly. *Angelica atropurpurea*, widespread across the Northeast, bears a strong resemblance to poisonous hemlock.

Fennel (*Foeniculum vulgare)*
Another garden escapee, fennel and bronze fennel can be found in parks and along boulevards, and often likes coastal locations. The leaves may be used fresh or dried—chopped, to impart the characteristic aniseed taste to a variety of dishes. The seeds that are gathered in the late fall can be dried and ground like peppercorns.

Marjoram (*Origanum majorana)*
Grows in grassy wastelands, and particularly on dry, alkaline soil. Use the leaves fresh or dried in the same way you would cultivated marjoram.

Wild herbs
1 Angelica
2 Fennel
3 Meadowsweet
4 Marjoram
5 Mint
6 Thyme

Meadowsweet (*Filipendula ulmaria)*
Grows abundantly in most locations, but particularly in damp, shady ground and near marshes. Both leaves and flowers may be dried and used for flavoring, or to make a refreshing, soothing drink.

Hang meadowsweet flowerheads upside down in a cool, airy place to dry, then you can crumble them into sweet dishes or preserves to add a subtle flavor.

Mint (*Mentha sp.)*
Various types of mint grow wild and it can spread quite aggressively in damp soil. Use whatever you can find in all the ways you use the mint you grow in your garden. Make a good supply of mint sauce to last you through the winter. Mint makes a most refreshing ice cream, too—with or without chocolate chips.

Sweet woodruff (*Galium odoratum)*
Found in woodland and along roadsides. The leaves may be picked and dried and are described as smelling like vanilla or honey. Can be used to flavor sausages and hamburgers and gives a pleasant taste to iced tea or lemonade.

Thyme (*Thymus sp.)*
Grows in grassland and open ground. It can be used just like cultivated thyme, although some say the flavor is milder.

Thyme is one of the most popular garden herbs, but also one of the most abundant found growing in the wild.

Fruit and Nuts

Wild berries grow all over North America; indigenous peoples ate them in a variety of ways: raw, dried, as fruit leathers, or with meat or fish. Never pick and eat berries unless you are sure you have identified them correctly.

Nuts have just as many culinary uses and, as an important source of protein, they can often be used as the principal ingredient of a meal. They also make superb additions to stuffings and sauces.

SOFT FRUIT

Pick and transport soft fruits carefully to avoid damaging them. Take a plastic tub with a lid, or line a basket with a plastic bag.

Barberry *(Berberis)*
The bright red berries appear in mid-summer. Beware of the thorns when picking. The name is also sometimes used for Oregon grape *(Mahonia)*. Use berries for jelly or as a sauce to go with meat.

Blackberry *(Rubus sp.)*
The most common of all wild fruits, and widespread in vacant lots, fields, roadsides, and woodlands. Native trailing blackberries may also be called dewberries. The familiar black fruits appear from late summer onwards, but they should not be picked after early fall, when they will have become soft and mushy. Blackberries can be used in jams, jellies, chutneys, desserts, and wine. They also freeze well. In the West and Northwest, the Himalayan blackberry has become a widespread and very vigorous weed.

Huckleberry *(Vaccinium sp.)*
Different species of *Vaccinium* are known as huckleberry, bilberry, and whortleberry, depending on the region. Check with a knowledgable local source to help identify edible huckleberries and bilberries where you live.

The berries are small, round, and blue-black in color, and they appear from midsummer to early fall. They make excellent jelly and wine or can be used in all types of dessert. Evergreen huckleberry (*V. ovatum*) makes a good berry plant in home gardens in the Northwest.

Cloudberry *(Rubus chamaemorus)*
Native to Scandinavia and northern Ontario these soft fruit are found in areas where the ground and atmosphere is damp. The fruits are similar to blackberries in shape but a pale, pinkish color and they grow on very low shrubs. If you can gather enough and they are ripe, they can be eaten raw as a dessert.

Cranberry *(Vaccinium macrocarpon)*
Native to marshy and boggy land in the Northeast. The principal, or traditional, use for the hard, dark red, shiny berries is as a sauce to accompany roast turkey and game. They also are delicious dried, and for use in baked goods such a muffins. Cranberries freeze very well.

Wild berries make rich and delicious compote to serve with yogurt or ice cream, or jams and jellies to enjoy all year round.

Elderberry *(Sambucus sp.)*
Both the flowers, which appear from early summer, and the berries, which follow them later in the season, are valuable crops. Elderberry shrubs and small trees grow in forests and can often be found in vacant lots. There are several species of blue and black elderberry that grow wild throughout North America; do not eat the red elderberry (*S. racemosa*) or its flowers.

The clusters of white flowers can be made into a drink by pouring boiling water on them (let it cool and chill it before drinking) or into wine. They impart a subtle flavor to other fruit puddings, jams, or jellies, and can also be dipped in batter and deep fried to make fritters.

The berries are ready for picking when they are dark and hanging heavily from the branches. They can be mixed with other fruit in jams, jellies, or desserts; made into sauces and chutneys; used to flavor spiced vinegar; or made into superb wine. Dried, they can be substituted for raisins in cakes and muffins.

Gooseberry *(Ribes sp.)*
The native American gooseberry is *R. hirtellum*, mostly found in the North and the West. Use them in any of the ways you would use their cultivated cousins, such as in tarts, sauces, jams, and wines.

Mountain ash *(Sorbus sp.)*
This tree may be found anywhere, particularly in cool woods. The heavy clusters of orange, yellow, or red berries appear from summer to fall and can be used with crabapples to make a jelly to accompany meat. They can also be made into wine.

Raspberry *(Rubus ideaeus)*
Wild raspberries grow in woods and clearings. The berries are ripe during summer. They can be used in all the ways suitable for cultivated fruit, but if you only find a few, use them to flavor vinegar for a salad dressing, or mix them with other fruit. Raspberries are delicate, so handle them carefully.

Rose hips steeped in boiling water make a soothing tea, or you can make them into a syrup rich in vitamin C (see below).

Redcurrant *(Ribes rubrum)*
This grows as a wild plant in the eastern United States as far south as Virginia. Look for it particularly along thickets or at the edge of a wood, near a stream. The round, bright red fruits appear in summer and may be used in fruit salads, or baked into pies. Redcurrant juice is a good addition to salad dressings.

Rosehips *(Rosa sp.)*
Many native roses grow across North America. The orangey-red berries appear from late summer until late fall. Rosehips are extremely high in

vitamin C. Some rosehips have hairy seeds that are difficult to digest, but the fruits can be made into syrup, tea, and wine. Rosehips are also used to flavor liquor such as gin, as well as cocktail bitters.

Strawberry *(Fragaria sp.)*
Wild strawberries grow in open forests, coastal areas, meadows, and on mountain slopes, depending on the species. The tiny fruits, similar to those of the alpine strawberry, ripen during summer and have a far sweeter flavor than any cultivated variety. They are rich in vitamin C. Ideally, eat them raw, with ice cream to make them go further or, better still, with champagne or liqueur poured over them.

HARD FRUITS

These fruits must all be cooked in some way before they can be eaten. They are not as sweet as most soft fruits, although a frost will reduce the sharpness of some, including sloes and plums.

Crabapple *(Malus sp.)*
The ancestor of all cultivated apples, it grows in a variety of habitats throughout the United States and Canada. The crabapple has been widely crossed with cultivated varieties of apples, and many of these have found their way back into the wild, so the crabapples you find might well be variable in color, size, and taste (true wild crabapples are incredibly sour). Pick them as you find them in summer and fall and make jellies, sauces, chutneys, or cider, according to their sweetness.

Medlar *(Mespilus sp.)*
It is uncommon to find a wild medlar tree. The American native Stern's

Crabapples cannot be eaten raw, like larger apples, but have many uses in preserves, jellies, and stewed puddings.

medlar was only discovered in Arkansas in 1990 and there are very few known specimens. The medlar found in Europe has brown fruits that are roughly pear-shaped. They make a good alternative to apple sauce to eat with pork, or may be roasted in the oven and eaten as an accompaniment to meat. They also make a delicious jelly.

Pawpaw *(Asimina triloba)*
Tropical-looking pawpaws are native to woodlands in eastern North America and may be found growing in thickets. The late-summer and fall fruits are greenish yellow and brown when ripe, with a flavor somewhat like bananas.

Persimmon *(Diospyros virginiana)*
Native American persimmons grow in the southeastern United States. (The fruit is not quite the same as the Asian persimmons typically found in supermarkets.) American persimmon fruit ripens after frost, and is very astringent until it has softened, at which point it becomes very sweet. You must separate the pulp from the numerous seeds before using the fruit in baking, typically for breads and muffins.

Sloe *(Prunus spinosa)*
The small, bluish-black berries are native to Europe and are the fruit of the thorny blackthorn. The berries eaten raw are extremely sour, although they can be mixed with apples to make a jelly, or even a fruit pie. The better-known use of sloes is to make sloe gin: prick the skins with a pin (or freeze them to split the skins and then defrost them), mix with an equal quantity of sugar—less if you do not want the gin too syrupy—and pack into bottles. Top up with gin and leave for two or three months. Strain off the liquor and either eat the berries or add them to an apple pie.

Wild plums *(Prunus sp.)*
The typical, small plum-shaped fruits of the tree may be red, purple, yellow, or green. There are many native plums all across North America; they may be found in habitats ranging from prairies to pastures to woodlands. They can also be used in pies, jams, jellies,

Scour the hedgerows for sloes in autumn—they are a popular wild crop and the bushes are often stripped by other foragers.

and sauces. Texas wild plums are often used to make wine. The bark, leaves, and seeds are toxic.

NUTS

Nuts fall from the tree when they are ready, but are protected from damage by their hard shells. If you find a nut tree, gather what you can from the surrounding ground.

Beech *(Fagus sp.)*
Nuts appear in early fall, although only once every three or four years. Four nuts are usually contained in the brown husk. They can be eaten but are usually made into oil. Grind in a coffee grinder, place in a muslin bag, and press with a heavy weight, letting the oil fall into a bowl.

Hazelnut *(Corylus sp.)*
Known also as cobnuts or filberts, these trees are found in woods and thickets and yield their tasty harvest mainly in early fall.

Sweet chestnut *(Castanea sativa)*
American chestnuts are almost extinct in their native habitat because of chestnut blight, but sweet chestnuts may be found planted as street trees. The prickly green cases, each containing up to three nuts, begin to fall off these tall trees in fall. Like beech nuts, they must be shelled and peeled, and they also have an inner skin that makes them very bitter unless removed. It is usual to cook them before eating them, or to make soups, stuffings, or additions to casseroles. Puréed, they can be used to make *marrons glacés* and other desserts.

Walnut *(Juglans sp.)*
There are a number of native walnut trees in North America, including black walnuts and butternuts. The nuts ripen and fall from the tree in mid-fall, although you can pick them earlier for pickling to go with cheese or cold meats. The ripe nuts can be used in many dishes, both sweet and savory. Spread them out to dry for several days first.

Wild nuts
1 Beech
2 Sweet chestnut
3 Walnut

Walnuts are delicious to eat as a snack, but also make wonderful additions to cakes, breads, stuffings, and salads.

Mushrooms

It is imperative to be able to identify mushrooms if you intend to gather them for eating, and the best way to learn to do this is to spend time collecting with someone already experienced. Contact your local mycological society or mushroom club.

Most mushrooms are found in their greatest numbers in forests where the ground is rich in leafmold and, therefore, humus. In general, they like warm and damp, but not water-logged, conditions. Edible species appear from spring through fall, depending on the type. For instance, porcini mushrooms are avilable for gathering in spring and fall, but morels are found only in early spring.

Meadow mushrooms

The most important exceptions to the forest habitat are the meadow mushroom and the horse mushroom, and these are among the most coveted and sought-after of all wild, edible fungi.

Both field and horse mushrooms are found in meadows and pastures. The horse mushroom, as its name suggests, likes grassy fields frequented by horses, but also cattle. Look for it near cowsheds or horses' shelters in fields and along riding trails. They often, although not always, appear year after year in the same spot.

Safety tips and rules for picking

In the Pacific Northwest and parts of California, commercial mushroom gathering has led to the imposition of collection limits either at the state level, or by local Parks departments, or the Provincial Forest Service. Your local mycological society will be aware of any restrictions on collecting in your area, so to avoid possible fines, check if there are such restrictions in place and collect mushrooms in a sustainable manner.

There are some simple, but essential, points to remember when collecting mushrooms. Always follow these rules to keep yourself safe and healthy—and above all, never pick or eat anything you cannot identify.

Harvest mushrooms by twisting them by the stalk so they break free at the base. Pulling them out of the ground destroys the whole plant and stops it growing back; cutting them with a knife will hamper your identification, as the base of the stem is often a guiding feature.

Collect them on a fine day, not when it is raining, as mushrooms will deteriorate quickly if picked when they are wet. Use waxed paper bags or put them into an open container—a shallow basket is ideal; a plastic bag is not, as it provides the perfect conditions for very quick decomposition.

The mushrooms you collect should be mature, but not so old that they are beginning to decay. Do not pick young mushrooms, the tops of which are still bunched or buttoned; a poisonous species could easily be mistaken for an edible one.

Pick only perfect specimens; not those that are ragged, torn, or slimy. Go through them again when you get home, and discard any that you feel are suspect. Cook or dry mushrooms (*see* page 209) as soon as possible after collecting—and wash them very thoroughly first.

A cluster of edible parasol mushrooms this large on the forest floor is a lucky find. Mushrooms can be found in the same spots year after year.

EDIBLE MUSHROOMS

The texture and flavor of all these edible fungi vary widely. In most cases, it is best to pick a medley of different sorts to use together.

Field mushroom (*Agaricus campestris*)
Found in damp grassland, often growing in a ring.

Horse mushroom (*Agaricus arvensis*)
Found in grassy pastures and fields grazed by horses, cattle, and sheep, and along riding trails.

Wood mushroom (*Agaricus silvicola*)
Found in damp woodlands.

Parasol mushroom (*Lepiota procera*)
Found at the edges of woods and in grassy clearings in woods or by roadsides. Pick as cap is beginning to open, and discard the stem.

Oyster mushroom (*Pleurotus ostreatus*)
Found growing on dead tree trunks, branches, or stumps, particularly hardwoods like oak, aspen, and poplar. Pick when young and stew slowly, or dry.

Chanterelle (*Cantharellus cibarius*)
Found in forests, on the soil, or growing from rotted trunks. Stew in milk; chanterelles need to be cooked slowly.

Morel (*Morchella esculenta*)
Found in thickets, coniferous and hardwood forests, and grassy banks. Also likes rich, bare, or burned soil.

Porcini or cep (*Boletus edulis*)
Found in coniferous forests; on pine trees in particular. Good for drying.

Bay boletus (*Boletus badius*)
Found in Northeastern forests, particularly coniferous.

Boletus erythropus
Found in forests, particularly coniferous, or in poor soil. Do not worry about the blue color to the flesh when cut or broken.

Porcini mushrooms are found in decidious and coniferous forests, especially around pine, hemlock, and spruce trees. They are good mushrooms for drying.

Shaggy ink cap (*Coprinus comatus*)
Found on roadsides, unpaved roads, and other places where organic matter is buried. Gather while gills are still white, discard stems, and scrape scales from caps.

Giant puffball (*Lycoperdon giganteum*)
Found in forests and grassy areas, including lawns. Eat when young and the flesh is white, not yellow, brown, or multi-colored.

Edible fungi

1 *Field mushroom*
2 *Horse mushroom*
3 *Parasol mushroom*
4 *Oyster mushroom*
5 *Morel*
6 *Cep*
7 *Shaggy ink cap*
8 *Cauliflower fungus*
9 *Common puffball*

Safety essentials

It is important to stress that you should never eat any mushroom unless you are absolutely certain it is an edible species. Never test by tasting; a remarkably small amount of a deadly species could kill you. And never risk an uncertain identification; if you are not completely confident, leave the specimen behind. Similarly, even when you know a species to be edible, if you are trying it for the first time, eat only a little; you might have a reaction to it, even if other people can eat it with no ill effects.

The common puffball (left) springs up from forest floors, through layers of decomposing leaves, but can also be found in grassland. It is edible, and best picked when the flesh is white.

Common puffball
(*Lycoperdon perlatum*)
Found in forests, grassland, and disturbed sites. Pick when flesh is white, not darkened.

Cauliflower fungus
(*Sparassis crispa*)
Found near Douglas fir trees, sometimes growing on the stumps or roots. Eat when young, or dry.

Hedgehog (*Hydnum repandum*)
Found in forests. Needs boiling for 20 minutes or so to prevent it from tasting bitter.

Blewit (*Tricholoma saevum*)
Found in pastures and other grassy areas, often in gardens, as well as in forests and thickets.

Wood blewit (*Tricholoma nudum*)
The same as blewit. Pick when underside and stems are tinged violet.

Yellow swamp russula
(*Russula claroflave*)
In wet ground under birch trees.

Bare-toothed brittlegill
(*Russula vesca*)
Found in forests, particularly of beech and oak.

Blackish-purple russula
(*Russula atropurpurea*)
Found in forests, mainly deciduous but also found in under conifers.

POISONOUS MUSHROOMS

Get to know these, so that you can avoid them at all costs.

Death cap (*Amanita phalloides*)
Found in forests, particularly oak and beech. It is deadly.

Destroying angel (*Amanita virosa*)
Found in forests; rarer than above.

Fly agaric (*Amanita muscaria*)
Found in forests or near coniferous and birch trees.

Panther cap (*Amanita pantherina*)
Found in forests, particularly beech, and sometimes grassy meadows.

Fool's mushroom (*Amanita verna*)
Found in forests, usually beech.

Brown roll-rim (*Paxillus involutus*)
Found in a variety of forest habitats.

Inocybe fastigiata
Found growing on forest floors.

Red-staining inocybe
(*Inocybe patouillardii*)
Found in forests and trails.

Clitocybe dealbata and ***C. cerrussata***
Found in coniferous woods.

Lepiota subincarnata Found in rich soil in parks and gardens.

Livid entoloma (*Entoloma simuatum*)
Found in forests, lawns, parks, and some gardens.

Sulphur tuft (*Hypholoma fasciculare*)
Found in large groups round the base of hardwood and coniferous trees. Can be found at any time, but usually during summer and fall.

Devil's boletus (*Boletus satanas*)
Found under beech trees in particular, but look under all trees that grow on alkaline soils.

Yellow-staining mushroom
(*Agaricus xanthodermus*)
Found in pastures, parks, and woodland.

POISONOUS FUNGI

1 Death cap
2 Destroying angel
3 Fly agaric
4 Panther cap
5 Inocybe fastigiata
6 Red-staining inocybe
7 Clitocybe dealbata
8 Clitocybe cerrussata
9 Devil's boletus
10 Paxillus involutus
11 Sulphur tuft
12 Livid entoloma
13 Yellow-staining mushroom

As sinister as its name, the death cap is one of the most deadly fungi and is found beneath oak and beech trees.

Fishing

Ocean and freshwater fishing in rivers and lakes are a fun way to boost your food supplies, though seldom a reliable source of fish throughout the year.

Always make sure you have the appropriate licenses and permits for wherever you want to fish. In most places, you cannot set up just anywhere with your fishing rod. Ask the advice of locals, check a nearby bait shop, or call a fishing charter company to find how who to contact for permission to fish.

OCEAN FISHING

For ocean fishing, you can fish from the shore or go out in a boat. Take care to follow the advice of the coast guard if bad weather is expected; small boats are very vulnerable in rough seas.

Fishing from the beach

You don't need a boat to catch fish from the sea. If you are patient, you can catch plenty of fish and shellfish without even leaving the shore and barely even getting your feet wet.

On a rocky beach try searching the tidepools at low tide, turning over the rocks one by one. If you are lucky you can get bucketfuls of clams, scallops, and mussels.

Another way of catching prawns and shrimp (which are mainly found on rocky coastlines) is by suspending a weighted cast net from the rocks into water that is about 4–5 ft (1.2–1.5 m) deep. Throw the net into the water so it reaches the bottom and leave it for 5–10 minutes. Pull it up and check inside for your catch. Bait the net with old fish if needed.

You may not be able to guarantee a catch, but fishing from a beach early in the morning on a good day will always bring you pleasure.

Using a long line

This is a method where a line is laid down the beach at low tide, the ends secured through a wooden board that is buried into the sand to anchor it. All the way along the line are short traces—6–9 in (15–25 cm) long—that have baited hooks placed on to the ends of them.

As the tide comes in, it will bring a number of fish in with it and they, hopefully, will take the bait. When you go back at the next low tide, there should be a line full of fish. If you live in an area where there are a lot of crabs, which will take your bait, put little corks on the traces. This will lift them as the water flows over them, taking them out of reach of the crabs.

Bait

All sorts of fish can be used for bait—an oily piece of mackerel, for example. Traditional baits include earthworms, night crawlers, leeches, waxworms, and minnows. The type of bait you choose will depend on the type of fish you are trying to catch. Many fish will not eat dead bait; others will not eat live bait. Speak with a bait shop in your area to find the best bait to use.

Mail-order companies will ship live bait if you do not have ready access in your region. However, some jurisdictions restrict what can be used as live bait, and also whether bait can be imported from another state or province.

Using a rod and line

If you have plenty of time to spare, try fishing with a rod and line from the beach, casting into the water at high tide. The length of line needed will depend on the sea bed and whether it drops away quickly, but on a normal beach you will need a line of at least 70 yards (64 m), so you will need to be an experienced angler.

Lobster pots

These basket-type pots are specially designed so that lobsters can swim into them, but cannot swim out. Although they are often put in 30 ft (10 m) of water or so from a boat, you can wade in and position them closer to shore around wrecks or rocks, or in estuaries where lobsters are known to be found. They should be weighted to keep them on the bottom and their position marked with buoys. Visit them periodically to remove anything you have caught.

Crab traps

There are many different designs of crab pots or traps, some of them are collapsible, so they are easy to transport. Like lobster pots, they are lowered into the water and left in place marked with a buoy. For crabbing off a dock, a crab ring—consisting of two metal rings with netting between them—can be the best type of trap.

Fishing from a boat

If you have a small boat, other possibilities will open up to you, but remember that it is unwise to go more than about 1 mile (1.6 km) offshore, as you may not get back to safety if the weather deteriorates quickly. You should only venture that far out in good weather conditions.

When considering offshore fishing, try to determine where the fish are going to be, so that you keep the chances of a fruitless mission to a minimum. Ask more experienced fishermen or learn by examining the sea bed at low tide. If there are gulleys or outcrops of rocks for example, these are the places where the fish will gather when they come in, so position your boat above them.

There are various techniques for fishing from a boat, and what you do will depend on the equipment you have and the type of fish you are trying to catch. Make sure your tackle is sufficiently heavy for the catch. Many ocean fish, like salmon, are strong and will put up a good fight before you can reel them in.

Offshore rod and line

If you are fishing with a rod and line, you will do better with bait than flies. If looking for bottom fish, anchor the boat and weight the line so the bait remains on the bottom. If you are going for the mid-water fish, let the boat drift, gently moving down the tide. You should play the line through the water: this means gently pulling it so the fish are attracted by the moving lures.

Your line can have several traces with hooks running off it, to maximize your chances. As soon as you feel a bite, haul the line in; if you have gone through a shoal of mackerel, for example, you may have a fish on each hook and if you are fast even with a small outboard motor, you can wheel the boat round and go through the shoal again.

Commercial lobster fishermen will drop large numbers of pots into deep water from a boat, but you can still catch crabs and lobsters by placing your traps in much shallower water, off a dock or closer to the shore.

If you are fishing with a rod, it pays to have a landing net, too, to help you land big fish without injuring them. You will also need a keepnet, a long net that you tether to a stake on the bank and where you can keep the fish you have caught while you continue fishing.

FRESHWATER FISH

How often freshwater fish will supplement your diet depends on where you live and whether you are likely to fish often enough to make it worthwhile obtaining a license.

Most rivers, lakes, and reservoirs cannot be fished officially unless you obtain a license to fish. In most instances these are issued by the local authorities, except in the case of privately-owned lakes, reservoirs, and stretches of river, in which case, permission has to be sought from the owner. Many national, state, and provincial parks also require licenses for fishing.

There are countless different species of freshwater fish, many of which are edible. The chief ones—and those most coveted by sporting anglers—are salmon, trout, and bass. In addition there are carp, chub, catfish, eel, lamprey, mullet, perch, pike, shad, sturgeon—and many more local species.

What will you catch?

This depends on where you are fishing and the time of year, as well as the day-to-day weather conditions, the temperature of the water, the spawning season, and so on. During winter, when the lakes freeze over, you can try your hand at ice fishing. Not all the fish you catch may be to your liking, but they are worth tasting at least once. Always check with the local authorities to make sure whether you need a license for whatever fish or other seafood you catch.

Trout fishing

Experienced trout fishermen can pull fish after fish out of the water without any kind of rod or net. One technique, known as "tickling" is to put your hand gently into the water (in a known trout river) and wiggle your fingers. As the trout drifts over your hand you continue the wiggling movement, tickling the belly, until, with one quick movement, you get your hand round it and either grab it around the body, or flick it out onto the riverbank.

Finding the best spots

You can increase your chances of catching fish by learning to recognize where they are. Bream, for example, tend to swim in shoals, and can be spotted by areas of muddied water in an otherwise clear stretch of river, while pike are lazy fish that often lie quietly in wait for their prey in the deepish water of a reed bed. If you do catch a pike, you can reduce its rather muddy taste by soaking it in vinegar and water.

Fish will be attracted to all sorts of bait—worms, maggots, garden slugs, and even bits of cheese, bread, and potatoes. Always check with the authorities where you plan to fish: some do not allow the use of live bait at all and you have to fish with artificial flies or lures.

Using a net

If you are fishing with a rod, it pays to have a net, too, to help you land big fish. You will also need a fishing bag or keepnet, a long net where you can keep the fish you have caught while you continue fishing.

Food from the Seashore

The shellfish, and the mollusks in particular, that live on the rocks or in the sand or tidepools at the water's edge are easy to gather, with no special equipment. You will also find many edible plants living close to the sea, which make the perfect accompaniment.

Pick your time and place for coastal foraging. Most shellfish are best looked for in tidepools and intertidal zones as low tide approaches. Once the water has ebbed away, they cling tightly to their rocks and are almost impossible to pry off.

State, provincial, federal, and local authorities all monitor shellfish to ensure that they are safe to catch and to ensure that they are not overharvested. Notices are usually posted in areas where there are known shellfish hazards, such as a rise in bacteria or algae levels in the water. However, before eating any shellfish, be sure you know it is safe and legal to harvest and eat in the area. If you are visiting a coastal area, look for recreational shellfish reserves if possible.

Mollusks

Shellfish, and mollusks in particular, should always be treated with caution but, by and large, they do not deserve the reputation they have for causing food poisoning. They feed by pumping water through their shells, filtering out the food particles as they do so, and any bacteria contained in the water tends to be retained. For this reason, mollusks found near sewage outlets should always be left alone, and it is wise to do the same with those living on piers, boardwalks, or other possible sources of pollution. Collect them only from clean, unpolluted stretches of water.

Some people advise against collecting mollusks during the warmer summer months. This is their breeding season, so they will necessarily not be in prime condition, and the warmer temperatures of the water can increase the chances of dangerous bacteria multiplying.

The other important point about shellfish of all kinds is that they decompose very quickly. Never collect any that are already dead (if the shell is open, or if they do not hold fast to their rocky stronghold, this is a sure sign). They should be alive at the moment of cooking, which should be done as quickly as possible after collection. Wash them very well first.

Clams

A great favorite, clams are most frequently found in the muddy sands exposed between tides. They are among the largest of the mollusks, and usually live quite deep beneath the surface, so you will have to dig for them. On the Atlantic coast, softshell steamers and hardshell quahogs are the most common species, followed by surf and razor clams. On the Pacific coast, clams vary from manila to Pacific littleneck to tiny, tender butterclams.

Clams are often difficult to pry open. Insert a sharp knife (preferably a special oyster-shucking knife) between the shells at the hinge and twist it. Cut off and discard the fleshy siphon. Alternatively, they can be persuaded to open by shaking the shells in a saucepan over a medium heat for a few minutes.

You can steam, grill, or fry clams. To steam them, place in less than an inch (2.5 cm) of water—seasoned with wine, lemon, garlic, and herbs—and bring to a boil. Simmer until the shells open; throw away those that don't open. You can eat them like this, fry them, or add them to chowder, stews, or pasta sauces.

Giant geoduck clams are found in the Pacific Northwest. They must be dug from the sand from a depth of up to 3 ft (1 m) at low tide. The meat can be used for stews, pasta sauces, or ceviche.

Limpets

These single-shelled mollusks cling to rocks, docks, piers, and other structures, but should be gathered only from clean rocks that are washed daily by the incoming tides. Pry them free with a knife and soak them before cooking. They are much tougher than clams and will need long, gentle simmering if they are not to be rather chewy.

Mussels

Northern blue mussels are found primarily on the east coast, though they can also be found on the Pacific coast, along with California mussels. These bivalves are found on rocks close to the shoreline. Because they are so susceptible to pollution, be especially sure that you only gather them from waters that are not under quarantine. Take them only from the low rocks that are washed by each tide. Discard any with broken shells, or those that do not close immediately if the shell is tapped.

Scrub the shells and soak them for five or six hours in cold water, preferably with a handful of oatmeal added. The mussels will feed on this, excreting the dirt in their shells. Discard any that open or float to the

Only gather mussels from places that are free from pollution. Discard any that are open or that open when soaked in water before cooking, as they are likely to be bad.

surface at this time then cook by steaming over a gentle heat or by baking them.

Oysters

There are many opinions on the relative merits of various east coast versus west coast oysters, but no matter what your preference, you will have to be aware of harvesting limitations and quarantines, especially for red tide (a type of naturally occuring algae bloom). If you do find oysters, do not let your enthusiasm mar your judgement—if they are in polluted waters, leave them alone.

Oysters live in shallow waters, often near or attached to rocks or stones. To eat them raw; pry open the shell open by inserting a sharp, strong knife at the hinge, twist it, then use it to free the oyster from the outer shell. Add a squeeze of lemon juice and swallow. Oysters can also be used in recipes for other shellfish.

Scallops

The edible portion of a scallop is actually the muscular tissue that holds the shell together. Sweet bay scallops are found in seagrass beds and can be harvested by wading through the shallow water with a net and a bucket.

In some places, scallops can only be harvested by diving or dip netting. Check with the local experts before you begin.

COASTAL PLANTS AND SEAWEED

Some coastal plants and seaweeds are edible and delicious. They are particularly well-suited to eating with the fish and shellfish found living alongside them.

Seaweeds are rich sources of iron and minerals. They are often treated with the same suspicion as mushrooms, but, like them, they can provide a tasty, free meal. However, as with any wild-gathered food, be sure you know what you are harvesting before you eat it.

Seaweeds are an essential ingredient of many Asian dishes. Always wash seaweeds very thoroughly in cold running water before cooking them; they are likely to be salty and gritty.

Seakale *(Crambe maritima)*
This plant grows wild in Europe and can be planted in coastal gardens. Pick off the big leaves, strip them to leave the stems, boil or steam these, and serve with melted butter.

Sea purslane
(Sesuvium portulacastrum)
This plant likes the salty marshes, dunes, beaches, and mangroves found along the east coast. The leaves may be used in salads.

Carragheen or Irish moss
(Chondrus crispus)
The reddish-purple fronds (which may turn green in very strong sunlight) of this seaweed can be found in shallow pools, clinging to rocks and stones. Gather it when young, wash well, and simmer slowly in the ratio of one part seaweed to three parts milk or water, adding sugar to taste. The seaweed will dissolve, at which point it can be strained and flavored to make a jelly, because the mixture will set firm when cold. Dry it by washing it and putting it in the sun. Use as gelatine.

Dulse *(Palmaria palmata)*
The fan-like fronds of dulse may be found in intertidal zones or hanging onto rocks on the shores of the Pacific Northwest and the North Atlantic seaboard. Dulse grows quickly in the summer. Discard the older parts, once gathered, as these tend to be tough; the younger parts are more tender and can be used raw in salads or cooked like a green vegetable. It can be pan fried, baked, or used in soups. Sun dried dulse may be ground to a powder.

Kelp *(Alaria marginata)*
Found round the low-tide mark on seashores, particularly those that are rocky. Kelp may be treated in the same way as carragheen to provide vegetable gelatine. It may also be used raw in salad, or dried and added to soups and stews, or deep fried and eaten like potato chips.

Laver *(Porphyra sp.)*
This is one of the most commonly-used seaweeds and its purply-green fronds, which turn black when dry, grow on all manner of stones and rocks. It particularly likes those that become covered with sand.

Sea lettuce *(Ulva fenestrata)*
The translucent green fronds of this seaweed will be found on all parts of the shoreline, sometimes floating in shallow water, sometimes in rock pools, and sometimes hanging onto stones or rocks. It can be washed and cooked like a green vegetable.

Low tide often reveals a rich harvest of coastal plants, such as sea lettuce (left) and other seaweeds. Always wash them well to remove salt and grit before use.

Hunting

There are many laws governing the killing of any wild bird or animal. There are also, of course, restrictions on gun ownership. Always ensure that you comply with all legal restrictions as well as make certain that you and any other people are safe at all times.

Pheasants are easy to shoot. Organized hunting trips—during the open season—are a good way to get some for the dinner table.

In the United States, hunting laws are usually a state jurisdiction, but there are also federal laws for certain species, such as migratory birds and endangered species. Most pest species can be hunted without a license (such as gophers, rats, and coyotes), but you should always check for local regulations. In Canada, hunting is similarly regulated at the provincial and federal level. You should never hunt on private property without obtaining permission from the owners.

Many jurisdictions divide animals into hunting categories such as game (deer, bear, cougar, moose, elk, and bighorn sheep). Predatory animals may include coyotes, rabbits, and rodents. Other animals may be protected or unprotected animals. Protected animals vary from one region to another and cannot be shot without a license during authorized seasons.

Unprotected animals many include rodents, certain birds, skunks, jackrabbits, oppossum, nutria, rats, and other small animals.

Rabbits

You will find rabbits in all wild habitats. If you catch a rabbit, check to see it is healthy before considering eating it. Rabbits should be paunched (de-gutted) as soon as possible after killing, or the meat will be tainted. Some hunters say they should not be hung like other game, but you may well find that the meat has a better flavor if you do hang it for a few days before cooking.

Game birds

As with animals, federal, state, and provincial authorities regulate bird hunting, including setting bag limits and open seasons.

Modern agricultural methods, with extensive use of chemical pesticides and herbicides, have depleted the supply of many game birds' favorite foods, so these birds are often now raised in captivity for hunting and food consumption, rather than existing in large numbers in the wild. These include pheasants, quail, partridge, squab, wild turkey, and grouse.

To tenderize the meat, game birds should be hung by the neck for about a week in a cool, well-ventilated place. In cold conditions they can safely be hung for longer. Although they are often hung with their feathers still on, they are best plucked as soon as possible after killing, as a warm bird is the easiest to pluck (*see* page 170 for instructions). The insides are left in during the hanging.

Gathering Firewood

Most woodland and forest belongs to somebody and, theoretically, you should ask the owner—whoever it might be—before you even pick up kindling from the forest floor.

If you are able to gather kindling, remember that it really means small bits of dead twigs and branches that have fallen to the ground. It does not mean snapping branches from the parent tree.

These small dead branches will start a fire quicker and better than firewood that has been chopped up small for kindling. If you have a local source, it is much quicker to go and gather an armful of twigs than to start chopping up old wooden pallets or splitting down large logs.

Be careful if collecting wood from old woodpiles or brush in areas that have poisonous snakes and spiders, as these are favorite hiding places for them. Likewise, if poison oak and ivy are prevalent where you live, wear protective clothing before you start gathering branches in the wild.

Felling trees for firewood

The only trees that you can chop down for firewood are those growing in your own property. Even then, you should check with your local or state authority that you are free to do so without contravening any tree protection order or other local bylaw.

If you are lucky enough to own your own woodlot, there could be a legal limit on the amount of timber you are allowed to cut down each year. Also you may need a harvesting license, although if you are felling fruit trees, or trees that are less than a certain height and trunk diameter, this may not be necessary.

FELLING A TREE

Felling a tree is simple if you know how and extremely dangerous if you do not. Seek the advice of experts and watch them at work before attempting it yourself.

The ideal situation is a perfectly erect tree standing somewhere where there is nothing to obstruct it as it falls. Before you start, it is vital that you plan which way you want the tree to fall and that you have planned an escape route from the site if the tree falls in the opposite direction. If you are working with a partner—which is a good idea if you

Always stack firewood in a dry, but airy place. Freshly felled wood should be allowed to dry, or "season," for up to a year before burning.

are felling many trees—keep each other in visual contact to prevent accidents.

Even if you know what you are doing, it is sensible to seek advice if the tree is in anything but perfectly straightforward circumstances. This would include if it is very large, leaning heavily, close to a deep ditch, utility wires, telephone poles, or other trees, or is growing on very uneven ground. Never attempt to use a chainsaw or other power saw without first getting proper instruction; these tools are extremely dangerous in inexperienced hands. Failing that, stick to the axe and hand saw. They may take longer, but they are generally safer. Always wear proper safety gear—steel-toe boots, goggles, and heavy gloves—to protect yourself when cutting wood.

If you cut down trees, you should also plant new ones to ensure future generations do not inherit a landscape denuded of trees. Plant young trees—the younger they are, the better they are likely to take and establish themselves. If you are planting hardwood, it will be 20 years before you can chop the new growth down for firewood.

HOW DO THEY BURN?

Different woods burn at different speeds and with varying amounts of heat and smoke.

For burning in an indoor fireplace or wood stove, use only fully seasoned hardwoods like red maple, sugar maple, hickory, oak, ash, beech, and locust. Trees like birch, aspen, willow, silver maple, pine, and larch are fast-growing, and so they have soft wood that burns quickly. In the West, Douglas fir, red alder, ash, and big leaf maple all make decent firewood. Old fruit trees, such as apple, plum, and pear, also make good burning.

All logs should be left to dry, or "season," before burning, otherwise they will cover the chimney with a black, gooey resin, which will soon close up the chimney altogether. The best way to do this is to stack them neatly in a covered shelter or inside a dry shed.

HOW TO FELL A TREE

1 Chop a large wedge in the tree on the side that you want it to fall towards.

2 Chop a similar-sized wedge slightly above the first, but on the opposite side of the tree.

3 Exert pressure on the same side of the tree to make it fall in the direction you planned.

4 Once the tree is felled, trim off the branches and chop the trunk into logs.

SPLITTING LOGS

1 Drive a wedge into the end of a log using a sledgehammer.

2 Drive the wedge in carefully so as not to splinter the wood.

3 Continue driving more wedges down the log until it splits in half.

4 Once the log is split, it can be chopped into smaller pieces as required.

PRESERVING YOUR PRODUCE

Some crops are prone to ripen in gluts—too many to eat all at once, but perfect to preserve for later in the year, by turning them into wines or chutneys, or storing in a cool, dry place.

Preserving Your Produce

Instead of buying out-of-season produce from the supermarket, gardeners can enjoy the preserved fruits of their own labor all year round.

Most foods taste best if they are harvested and eaten immediately, but there are few places where the climate allows for year-round growing and harvesting. For most of us, canning, drying, smoking, or otherwise preserving food is part of being self-sufficient.

In all preserving, it is crucial that you carefully follow any instructions—especially the cooking temperatures and times—and, in most cases, use only perfect produce at the start. If there is any question of the safety and reliability of a preserved food, throw it away.

It is a good idea to keep a record of each batch of produce preserved, recording what it was, how much it weighed, how long it took, the shelf life, and so on. You will soon develop a rhythm of planting, harvesting, and preserving that will fit the flow of the seasons and the harvest from your own land.

Fruit and vegetables

The methods of storing and preserving depend on the individual fruit or vegetable. Many can be frozen. Some, like root vegetables, are easy to store in cool, dry places. Another popular method of preserving vegetables and herbs is to keep them in vinegar to create pickles, chutneys, relishes, sauces, and ketchups.

Freezing fruit—soft fruit in particular—will alter its texture; raspberries are often mushy when defrosted, but can still be delicious made into a sauce or pie filling. Some fruits and vegetables can be dried and used in that form or after being rehydrated in water. Apples and tomatoes dry especially well when cut into thin slices. Fruit is sometimes better preserved in syrups or purées.

Making wine, beer, and cider at home saves money and uses up excess fruit, but requires patience if it is to be worth drinking.

Meat and fish

If you keep goats, pigs, or chickens, you may wish to preserve the flesh. Salting and smoking are useful techniques, but the meat will not keep for very long. You can also preserve the products that come from your chickens and goats. Eggs can be pickled for storage and goats' milk turned into a range of dairy products, such as yogurt and soft cheeses.

Preserving Vegetables

Vegetables can be stored and preserved in a number of ways. Some leave the vegetable in its natural state so that it can be used as if it had just been harvested; others change its nature, but still allow you to enjoy your home-grown produce, whatever the season.

Root vegetables store most successfully in their natural state, and they can also be frozen. Freezing is probably the most successful method of preserving most other types of vegetables in order to keep them as close as possible to their natural state.

A few vegetables, such as herbs and greens (like kale or endive), have too high a water content for successful freezing. They can really only be stored when you cook them first, such as making stew or soup, which you then freeze.

LEAVING IN THE GROUND

All root vegetables, and some others that have a very long harvesting season (such as leeks), can be left in the ground through the winter and dug up as you want them.

There are three main objections to this as a storage method: you are unable to use the ground for anything else; the bed may be so hard that you cannot get a shovel into it; and it can be very cold work—picking Brussels sprouts when they are frozen to the plant and the frost or snow is thick on the ground is few people's idea of fun. A compromise to storing them in the ground is to dig up the root vegetables and bury them again in large pots of damp peat or potting soil. Store the container in a frost-free place over winter, keeping the soil just moist.

TRENCHING

This old-fashioned way of storing potatoes and root vegetables is still effective and can make use of a unused piece of ground.

Vegetables like cabbage—which has a strong odor that can penetrate an indoor storage space—can be stored in a mound or a trench outdoors. Dig the cabbage with their roots intact. Dig a shallow trench and place the cabbage head-down in the trench. Cover with straw and soil.

STORING INDOORS

It is often easier to store root vegetables in a dry, cool, frost-proof place. Always remember not to wash them before storing, or they will rot.

The main requirements of a root cellar or winter storage space are that it be dry, cool (or at least not hot), and protected from insects and rodents. Beware of storing vegetables in a garage that is in regular use for keeping cars or gas-operated lawn and garden equipment, as the vegetables could easily take up the taste of gas fumes.

Some vegetables prefer to be in slightly damp sand, peat, or sawdust. Beets, carrots, celeriac, celery,

Root vegetables will keep well if you bury them in large pots of damp peat or soil, stored in a frost-free garage, basement, or shed.

Jerusalem artichokes, rutabagas, and turnips prefer cool, slightly moist places.You can layer them in containers, separated by sand or peat, which prevents the vegetables from drying out or shriveling.

Use wooden boxes—old-fashioned fruit-picking trays or wine boxes are ideal—or pile the vegetables in white plastic buckets and put a 2 in (5 cm) layer of sand in the bottom. Put a single layer of the vegetable on top, arranged neatly and close together. Top with another layer of sand, and then continue layering in this way.

The option that requires least effort is to pile up vegetables that prefer dry conditions—potatoes, onions, garlic, winter squash, and dried beans—in baskets or boxes in the storage space and use them only as required. Of course, they are vulnerable like this to mice or any other rodents.

Storing in sacks

Most root vegetables can be stored in heavy paper or burlap sacks, but not plastic, which holds in the moisture and will make them rot. Storing in sacks is a good way of keeping potatoes; pile them into the sack, leave it open for a couple of days, then close it up and store in a storage space as above. Potatoes must be kept away from the light or they will turn green. Sweet potatoes must be cured in a warm dark place for up to two weeks before storing; they will not store for longer than a month or so.

Stacking in boxes

Fully ripe onions, shallots, and garlic can be stored in shallow slatted wooden boxes. Put them into the boxes in single layers and stack the boxes one on top of another.

Loosely pack root vegetables, such as these beets, into wooden trays and cover with sand to keep them slightly moist and fresh until you are ready to use them.

Tying onions in strings

The other traditional way of storing onions is in strings or ropes, and these should again be kept somewhere cool and frost free. There are various methods of stringing the onions together, but always check first that the onions have been dried ready for storage and have long stalks.

A simple method of tying onions in strings is to knot the stalks of four onions together and braid a strong piece of string around them, so that they hang neatly. As you add each new onion to the collection, knot the stalks around the string, making sure that the bunch always hangs evenly. When you have strung enough onions, hang up the bunch in a cool, dry place.

Using nets

Onions can also be hung up in nets, but check them frequently to make sure none are rotting and infecting others. This is also a way of storing winter squash or pumpkins.

Pumpkins can be stored on a shelf in the shed, as can winter squash, but they will not keep very long. Cauliflowers can be kept for a few weeks if hung upside down by their roots in a cool shed.

DRYING

A few vegetables can be dried, although this is more frequently done with herbs (*see* page 219) or fruit. Store dried vegetables in airtight jars or plastic storage bags.

Beans, peas, mushrooms, tomatoes, and onions can all be dried. The job is most easily and safely done in a dehydrator, which can be found through mail-order suppliers, or sometimes in natural food stores. They consist of a double-wall construction of metal or plastic, a number of removable mesh trays, and a thermostat with temperatures up to 160°F (70°C). If you do not have a dehydrator, use your oven set at a low temperature—between 100°F and 160°F (70°C). If you live in the desert, you can dry vegetables outdoors. Place on trays and cover with cheesecloth or row covers to keep off insects. Bring in at night.

Pole, bush, and runner beans

Pick beans when young, and top, tail, and string them if necessary. Wash the beans and slice the runner beans. Plunge into boiling water for about three minutes, rinse quickly in cold water, then spread them out on a clean tea towel or a thick wad of paper towel.

When they have dried a little, spread them out on the dehydrator trays and leave until they are dry and crisp. They should be soaked for several hours in cold water before being cooked in the usual way.

Peas

For shelling peas, pick and shell the pods. Spread the peas on the trays and dry in the dehydrator until brittle. For snow peas and snap peas, follow the method for beans. Drying on the plant is usually less successful, unless you grow a variety specifically bred for this.

Tomatoes

Dried tomatoes have an amazing sweetness and can be used in salads and sandwiches, or mixed into pasta sauces. Roma, cherry, and pear tomatoes work best for drying.

Dunk tomatoes in a pot of boiling water to loosen the skins, then peel them. Slice or quarter the fruit and arrange on dehydrator trays. Dry until crisp.

Mushrooms

You can dry those you collect from the wild, but they must be absolutely fresh and the very best specimens. Peel them if they look dirty, otherwise wipe them clean with a damp cloth or brush with a mushroom brush. Remove the stalks put them in a single layer on dehydrator trays, then dry until brittle.

An alternative method is to thread the caps onto a piece of string (making sure they do not touch each other). Hang strings in a warm place (such as above a water heater or above the stove).

Leave until they are dry and crisp and can easily be crumbled, then store in an airtight jar. For frying or grilling, boil them in a little water for about 15 minutes, or soak for an hour or two. They can be added as they are to soups, pasta sauces, and stews along with other ingredients.

Onions

Peel and cut into ¼ in (0.5 cm) slices. Separate into rings and plunge into boiling water for 30 seconds. Drain and rinse quickly under cold water. Drain on a tea towel and then spread on the dehydrator trays. Dry until they are crisp. Soak the onions for 30 minutes before using them.

BRINING

This is an old method of curing beans, cabbage, cauliflower, and pickling cucumbers. Essentially the vegetables are fermented in a liquid brine—a mixture of pickling salt and water. The vegetables can then be used to make sour or sweet pickles.

Buy pickling salt, which has a finer texture and so distributes evenly through the brine. A good 10 percent brine solution is 1 lb (450 grams) of pickling salt to 1 gallon (3.5 liters) of water. Weigh down the vegetables to keep them fully immersed. After a day, add more pickling salt at about ½ lb (225g) for each 5 lbs (2.3kg) of vegetables. For the next five weeks, add 5 tablespoons of salt for each 5 lbs (2.3kg) of vegetables. The vegetables will ferment in the brine for up to eight weeks. Remove the scum that forms on the top.

Preparing vegetables for brining

Pick the beans when young, top and tail them, then string, wash, and slice them. Blanch beans in boiling water for 5 minutes. Cool, then brine. Slice cucumbers fairly thinly or leave small gherkins intact. Slice cabbage, and cut cauliflowers into florets before brining. After brining, vegetables should be desalted before pickling by soaking for one day in fresh water.

Stringing onions and hanging them is the traditional way to store this vegetable. Keep long stalks on the onions when you lift them.

Freezing tomatoes

Tomatoes can be frozen once they are ripe, although they can only be used for sauces, soups, or stews once frozen; they are no good for salads because they are too mushy once defrosted. You can also store green tomatoes harvested at the end of the season by wrapping them individually in newspaper and keeping them indoors in a drawer or box. They should keep during fall right up until the middle of winter.

FREEZING

This is probably the most successful method of storing vegetables, but it can be tempting to keep more than you could ever use. Freeze only as much as you need to last you until the next fresh harvest; frozen food does not last indefinitely.

Vegetables should be frozen as soon as possible after picking, so only pick at one time the amount you can reasonably process right away. If you intend to freeze a sizeable proportion of any one type of vegetable, it is sensible to grow one of the varieties that have been specially developed for freezing. Only freeze vegetables that are in prime condition; discard any that are bruised or damaged.

Pack vegetables for the freezer in meal-size quantities. This is less important for those vegetables that you "open-freeze," laying them out individually on trays and freezing before packing into bags and containers, as they remain separate when frozen and may be tipped individually from the bag. Open-freezing takes a little more time, but produce treated this way is easier to handle than solid blocks of vegetables in ice. It is suitable for peas, beans, zucchini, baby carrots, and other vegetables that grow in small individual pieces.

Blanching

It is usually best to blanch vegetables before freezing them, as this helps to retain the color, flavor, and texture of the vegetable and also preserves the vitamin content. A stainless-steel blanching basket is a worthwhile investment and the prepared vegetables are then plunged into boiling water for the prescribed time (*see* opposite). After this they are plunged into ice-cold water for the same length of time to stop the cooking process.

Use a large pan of water when blanching, as it is essential that the water returns to a boil within one minute when the vegetables are immersed in it. The recommended amount is 6 pints of water to 1lb of vegetables (2.8 liters to 450 g). Don't try to put more than this amount of vegetables into the blanching basket at any one time.

The same water can be used to blanch about six batches of vegetables, but replace the cooling water each time, as it will have warmed up slightly. Fill a bowl with cold water and place some ice cubes into it. Drain the vegetables well after cooling, and pat them dry with paper towels before packing, labeling, and freezing them.

Frozen food safety

Follow the normal freezer rules of packing vegetables into rigid plastic containers or plastic freezer bags, excluding as much air as possible in both cases. Label them with their name and the date they were frozen and then keep a record so you can use produce in rotation. Always leave some "headspace" between frozen foods in the freezer because foods expand when frozen.

What can you freeze?

The chart opposite gives instructions for preparing vegetables for freezing, and the length of time they should be blanched in boiling salted water, unless otherwise stated. Not all those vegetables included in the growing section of this book will be found on the chart, because some do not freeze well. Most leafy greens have too high a water content to freeze, and some root vegetables—such as Jerusalem artichokes or rutabagas—can only be frozen as a purée. The exception to this is if you want to freeze a small selection of mixed vegetables—including a variety of roots—to add to stews or casseroles. In this case, prepare and blanch all the different vegetables individually and then freeze them together in a bag.

Beets and red cabbage can be frozen, but are usually preserved by pickling (or, in the case of beets, in boxes of sand). Cucumbers, endive, tomatoes, and radishes do not freeze well because of their high water content.

Freeze vegetables in meal-sized portions, in bags or boxes, or open-freeze them on trays and put into bags once they are frozen.

FREEZING VEGETABLES

VEGETABLE	PREPARATION	BLANCHING TIME
Artichokes	Cut off stems and coarse outer leaves. Remove the choke if you wish; wash thoroughly in cold water	7 minutes in water with lemon juice added. Blanch 5 at a time
Asparagus	Cut off the woody, lower stems and wash spears well. Sort into various thicknesses	2 minutes for thin stems 4 minutes for thick stems
Beans, broad	Choose young beans and shell them. Can be open-frozen	2 minutes
Beans, bush and pole	Wash and cut off the ends. Slice or cut into 1 in (2.5 cm) lengths	2 minutes
Broccoli and broccoli raab	Choose small, tender shoots. Cut off tough stalks and leaves, wash thoroughly in salted water, and grade into thick and thin stems	3 minutes for thin stems 4 minutes for thick stems
Brussels sprouts	Choose small, firm sprouts; remove outer leaves and wash in salted water. Grade into sizes if they vary. Can be open-frozen	3 minutes (4 if they are big)
Cabbage	Choose a firm-headed variety that is young and crisp. Shred and wash thoroughly in salted water	2 minutes
Carrot	Young carrots are best. Wash (rub off the skins after blanching) and cut off tops. Leave whole and open-freeze. Wash, scrape, and cut larger carrots into slices	4 minutes for whole carrots 3 minutes for sliced carrots
Cauliflower	Choose compact, white cauliflowers and cut into small florets. Wash well. Can be open-frozen	3 minutes in water with lemon juice added
Celeriac	(If wanted for cooking, not for eating raw) Peel and cut into cubes	4 minutes
Celery	(If wanted for cooking, not for eating raw) Choose crisp young stalks. Cut off roots and leaves, wash thoroughly, and cut into even-sized lengths	3 minutes in water with lemon juice added
Corn	Choose young cobs. Remove the green husks and silky tassels and sort into sizes. If you like you can cut the kernels from the cobs and freeze them	4 minutes for small cobs 6-8 minutes for larger ones 3 minutes for kernels only
Eggplant	Wash and cut into ¼ in (0.5 cm) slices. Drop these into water with lemon juice added to avoid discoloration. Can be open-frozen	4 minutes
Kale	Choose young shoots, trim off thick stalks and older leaves. Wash thoroughly	3 minutes, but not essential
Kohlrabi	(If wanted for cooking) Choose young ones, peel, and cut into even-sized chunks	6 minutes, until almost tender
Leek	Cut off roots and green top. Remove outer leaves. Wash thoroughly; leave thin ones whole and split thicker ones into even-sized lengths. Can be open-frozen	3 minutes
Okra	Wash and remove stems, without exposing seeds	4 minutes
Parsnip	Choose young, unblemished parsnips. Top and tail, peel, and cut into chunks	2 minutes
Peas, shelling	Choose young ones. Pod and grade into sizes if they vary. Can be open-frozen	1 minute
Peas, snap and snow	Wash peas and leave intact	3 minutes
Pepper	Wash, slice, and remove seeds and white membrane. Can be open-frozen	6 minutes, until almost tender
Potato (new)	Choose small, even-sized potatoes and scrape or scrub them	6 minutes, until almost tender
Salsify	Choose young roots, peel, and cut into even-sized lengths	2 minutes in water with lemon juice added
Spinach	Use young, unblemished leaves and process immediately after picking. Strip leaves from the stalks and wash thoroughly	2 minutes. Drain well and squeeze out excess water
Squash, summer and winter	Choose young, firm ones. Peel, discard seeds, and cut the flesh into even-sized, fairly large chunks	2 minutes
Swiss chard	Choose those with tender midribs. Remove leaves (and use or freeze like spinach—*see* below) and cut ribs into even-sized lengths	3 minutes
Tomato	(For use in stews, soups, etc.) Choose firm tomatoes and wash and dry them. Freeze whole	No blanching
Turnip	Choose young, small turnips. Cut off the ends, peel, and cut into chunks. Small ones can be left whole	3 minutes (4 minutes if freezing whole)
Zucchini	Choose young, firm zucchini. Trim ends, wash, and cut in half lengthwise or slices. Can be open-frozen	1 minute

Equipment for Making Preserves

Most of the things you will need for making sweet or savory preserves will probably already be in your kitchen, but they must all be made of materials that will not react with the acids in fruits and vinegars. Good choices include stainless steel, glass, plastic, nylon, or wood.

To make preserves, you can buy a special jam-making pan, called a maslin, which is very large and teacup-shaped, with a sturdy carrying handle and a lip for pouring. But as long as you have a standard large saucepan with a heavy bottom, you can make preserves. The pan should be stainless steel or have a non-stick or enamel lining. When cooking sweetened mixtures, keep the pan no more than half full—the mixture will spit when boiled—so adjust the quantities of a recipe to suit the size of your pan.

You will also need the following:

Scales Most preserves can be made without precisely measuring the ingredients, but some scales can also be helpful. For making large quantities, you may find bathroom scales a better choice.

Knives Always use sharp, stainless-steel knives to make clean cuts and minimize the discoloration of fruit.

Ladle For transferring the preserves into their jars, a ladle is far easier than trying to tip the heavy pan.

Spice bag A small square of muslin can be tied around a collection of spices, or you could use coffee filters. Special drawstring bags or metal spice balls are also available.

Spoons A long-handled wooden spoon is essential for frequent stirring without burning yourself. Scum is often produced when making jams and jellies; skim this off with a slotted spoon to prevent the finished preserve from turning cloudy. You can also use a slotted spoon for removing fruit seeds and pits during cooking.

Thermometer You can test whether jams and jellies have reached their setting point by dropping a little onto a cooled saucer, but a sugar thermometer designed for measuring high temperatures is more precise.

Sieves and **colanders** Make sure that they are either stainless steel or nylon.

Bowls and **measuring spoons** or **cups** You will need a selection of bowls, jugs, and measuring equipment.

Funnel These make it easier and safer to pour hot preserves into jars and bottles. Wide-necked funnels are available for jars and standard narrow ones for bottles.

Jelly bag and stand Jellies must be strained through a mesh bag part-way through cooking. A bag and strainer stand is a good investment.

Pressure cooker This is not essential but will save a lot of time for recipes where you need to pre-cook vegetables to soften them.

JARS AND BOTTLES

You can buy mason-type jars new in standard sizes or save used jars and wash them for your own preserves. You will need to buy new self-sealing lids. Wide-mouthed jars are most useful for pickles, as they are easier to fill.

Always inspect jars carefully before use to make sure that there are no cracks, chips, or other damage. If you pour hot jam into a flawed jar it will shatter; even if it does not, the cracks will harbor bacteria that could spoil your preserve.

Sterilizing jars

Wash jars in hot soapy water then rinse in hot water and stand them, upright, in a large, deep pan. Fill the pan and the jars with boiling water and boil rapidly for 10 minutes. Use tongs to lift the jars out of the water and drain them upside down. Use the jars warm if the preserve you are putting in them is also warm, but allow them to cool before filling with a cold preserve.

Always fill jars to just below the rim; if you leave a lot of room at the top, the air trapped inside the jar can contain microorganisms that will spoil the preserve.

Covers and lids

Acid-proof lids can be used for sweet preserves, but vinegary chutneys and pickles must be sealed with a vinegar-proof lining. If you only have metal lids, which are not vinegar-proof, cover the jar with plastic wrap before putting on the lid.

You don't need much special equipment, but may find it useful to buy a jam thermometer (top left), jelly-straining bag (top right), or piece of muslin (below right) and a ladle (below left).

Pickles, Chutney, and Relishes

Many vegetables can be preserved in vinegar by making them into pickles, chutneys, relishes, sauces, or ketchups. Spices add a tangy and often sharp flavor, and the resulting preserves are delicious as condiments to many cooked or raw dishes, or stirred into soups, stews, stir-fries, and more during cooking.

The most important thing to remember in the preparation of all these preserves is that any cooking must be done in an aluminum, stainless-steel, or enamel-lined pan, because vinegar will react with copper or brass and taint the taste of the preserve. Use wooden

Preserving will help to prevent produce that tend to ripen in a glut—like zucchini and tomatoes—from going to waste.

CHOOSING A PRESERVING METHOD

PRESERVE	METHOD	SUITABLE VEGETABLES
Pickles	Pickling retains the shape, color, and texture of the vegetable, while preserving it in spiced vinegar. Many fruits can also be pickled, as well as hard-boiled eggs	Beans (pole or bush), beets, cabbage (red or white), carrot, cauliflower, celery, cucumber, mushroom, onion, pepper, tomato (red or green), and zucchini
Chutneys	Chutneys are preserves in which vegetables are chopped and cooked slowly with vinegar, spices, and sugar, and often with fruits such as apple, dates, and raisins. They have a jamlike consistency	Beets, beans, squash, pepper (red and green), pumpkin, and tomato (red and green). Good fruits for chutney include apple, apricot, blackberry, damson, gooseberry, orange, pear, and rhubarb
Relishes	Relishes retain the crispness of the ingredients by chopping the vegetables coarsely and cooking them quickly or not at all. Most relishes should be kept for around two months before using	Tomato, pepper, and celery; corn, pepper, and onion; cucumber and onion; beets and cabbage; other combinations according to what is available
Sauces and ketchups	Sauces and ketchups are made by cooking vegetables with vinegar then straining and cooking again to give a concentrated flavor and smooth consistency. Sauces may be made from several different vegetables, but ketchups are usually made from only one	Mushroom, onion, and tomato

Above left *Make sweet pickles, such as figs and lavender, by adding sugar to the vinegar.* **Above center** *Pickled cucumber makes a tangy accompaniment to cold meats.* **Above right** *Add lemon or herbs to give a fresh flavor that will complement summer vegetables, such as these pickled artichokes.*

spoons and stainless-steel knives, rather than metal ones, and nylon or stainless-steel sieves. Vinegar will corrode metal jar tops unless you buy specially-treated vinegar-proof ones. Jars with airtight plastic screwtops or clip-on lids with thick rubber gaskets are good alternatives, but always ensure that jars are air-tight to prevent the vinegar from evaporating.

MAKING SPICED VINEGAR

You can buy ready-spiced pickling vinegar, but it is fun to make your own from a good basic malt vinegar so you can tailor the flavors and level of spice to suit your taste.

The most common spices used are cinnamon, allspice, cloves, chilis, and peppercorns. Always use whole

PICKLING VEGETABLES

Choose young, unblemished vegetables for pickling and make sure that they are crisp and fresh, as they will be preserved just as they start. Save less-than-perfect specimens for chutneys. If the vegetables are cooked before pickling, pack them into warm jars while they are still hot; raw vegetables should be packed into cold jars.

1

1 Wash and chop the vegetables if necessary (some, such as onions, are best pickled whole). Steep the vegetables in coarse salt or salt water for 24 hours to draw out any water they contain, which would dilute the vinegar.

2

3

2 Wash the vegetables well then drain them thoroughly. Pack them into clean jars up to 1 in (2.5 cm) from the top.

3 Drain off any water that collects in the jar then fill with the strained spiced vinegar to ½ in (1 cm) from the top. Seal the jars immediately with airtight lids.

Unopened jars can be stored for 18 months in a cool, dark place. Once opened, keep in the fridge and use within a month.

spices, as powdered ones will make the vinegar cloudy. Add about ¼ oz (6 g) of each per 2 pints (1 liter) of vinegar. Use more spices for a stronger flavor and add bruised ginger, mustard seeds, and dried red pepper for heat or sugar for a sweet pickle. Leave the spices to steep for at least one month. If you want a light-colored pickle, start with a white, distilled vinegar.

CHUTNEYS

For a chutney, the vegetables are prepared by chopping them finely or mincing them to ensure the finished chutney has a smooth, even texture—how finely you chop will depend on your personal taste.

The vegetables are cooked slowly with vinegar, spices, sugar, and any other additions (such as fruit or raisins) for a long time. Malt vinegar is usually used, although white, distilled vinegar, red wine vinegar, or balsamic vinegar can also be used. It may be spiced as for pickles or the spices may be added during the cooking—ground spices can be added directly, but whole spices should be tied in a spice bag, suspended in the mixture throughout cooking and then removed.

The spices used are similar to those used to spice the vinegar in pickle-making. Individual recipes will specify their own combinations. Most recipes include sugar, and brown sugar generally gives the best color.

Achieving a soft texture

If vegetables are particularly tough (onions for example) it is often advisable to soften them first by cooking them in a small amount of water or vinegar in a covered pan. Once they are soft, add the rest of the ingredients and the remaining vinegar and continue the cooking with the pan uncovered: the liquid must be able to evaporate to produce the desired pulpy texture.

Chutneys generally require a good one or two hours' cooking, by which time there should be no liquid left on the top of the vegetables or round the edges of the pan. If you think the mixture has become too solid (and remember it will thicken still further as it cools), add a little more vinegar.

The hot chutney should be poured into warm, clean jars and covered immediately with airtight lids. The flavor will mature and mellow during storage, and all chutneys should be kept for a few months before they are used.

MAKING CHUTNEY

It is not so important to choose crisp, young vegetables as it is when pickling produce, but make sure all blemished parts of those used are cut away and discarded.

1 Cut the vegetables and fruit into small pieces. Salt watery vegetables, such as zucchini, and leave them to drain in a colander overnight or for several hours at least before rinsing and ready for use.

2 Put all the prepared fruit and vegetables into a large pan together with the vinegar and spices. Bring to a boil and simmer, without a lid, for between 30 minutes and 1½ hours, or until the mixture is soft, but not mushy. Stir occasionally to prevent sticking.

3 Turn the heat to low and stir in the sugar. Once it has dissolved, return the pan to a boil and cook until the chutney is thick, stirring regularly to prevent sticking. The chutney is ready when no liquid appears when you drag a spoon across the bottom of the pan. It will thicken more as it cools.

4 Spoon carefully into jars and seal them promptly.

Making relishes

As with pickles, vegetables should be fresh, crisp, and young. Wash and chop them into small, coarse pieces and cook as directed in the recipe, although for many relishes the vegetables are left raw. Add spiced vinegar and, usually, sugar and allow the relish to mature for around two months before use.

SAUCES AND KETCHUPS

These are prepared in the same way, but ketchups usually only have one vegetable ingredient, while sauces may include several.

Only use vegetables that are ripe and well-flavored, and discard any bruised or damaged parts. The vegetables should then be washed, and chopped finely or minced before being cooked together with the vinegar and spices as stipulated in the individual recipe. When the vegetables are soft and pulpy, push them through a sieve and return to a clean pan to continue cooking.

If sugar is to be used in the recipe, it is generally added after the vegetables have been sieved, and additional vinegar may be added at this time too.

If using corks to seal your bottles, let them steep in boiling water for about 15 minutes first. This sterilizes and also softens them, making it easier to push them into the bottles.

Sterilizing the sauce

Sauces and ketchups made from vegetables that have a low acid content—tomatoes and mushrooms in particular—may ferment during storage, so they have to be sterilized. Screw the caps loosely, or tie down the corks, otherwise they will blow out during the sterilizing process.

Put a false bottom in a pan that is deep enough to take the bottles, so that water can be added to come up to the bottom of the screw tops or corks. The false bottom can be made from thickly folded newspaper or by using a piece of slatted wood or a wooden or metal trivet. Put the bottles on top of this, making sure they do not touch each other or the sides of the pan.

Pour in sufficient warm water to come up to the bottom of the corks or bottle caps and heat the water to a temperature of 170°F (77°C).

Maintain the temperature of the water for 30 minutes, then remove the bottles, using tongs, and screw them up tightly. These sauces and ketchups should be used quickly once they have been opened. Those that do not require sterilizing will keep for several months once opened, as will pickles, chutneys, and relishes.

MAKING TOMATO KETCHUP

Once you have tasted home-made ketchup, you will never go back to the sweet, but bland, commercial varieties in the shops. Add a couple of teaspoons of sun-dried tomato paste to boost the flavor if your tomatoes are not quite ripe. This makes enough for one bottle, about 2½ cups.

INGREDIENTS

- 3 lbs (1.3 kg) ripe tomatoes
- 2 red onions
- 2 celery stalks
- 2 cloves garlic
- 7½ oz (220 g) light brown sugar
- 6 floz (175 ml) cup cider vinegar
- ½ teaspoon cayenne pepper
- 1½ teaspoons paprika
- ½ teaspoon salt
- 2 teaspoons sun-dried tomato purée if desired

1 Chop the tomatoes, red onions, celery stalks, and cloves of garlic and put them all in a large pan. Cook over a very low heat until the tomatoes are pulpy.

2 Increase the heat and boil rapidly until the mixture thickens, stirring frequently to prevent it sticking or burning on the bottom of the pan.

3 Press the mixture through a sieve and return it to the cleaned-out pan or a fresh one. Over a low heat, stir in the remaining ingredients.

4 Stir continuously until the sugar has dissolved then simmer the ketchup, stirring frequently, until the mixture has thickened to the right consistency.

5 Bottle and then sterilize the ketchup (*see* below). Store in a cool, dark, dry place for one month before using, but do not keep for longer than one year.

STORING AND PRESERVING VEGETABLES

The most successful and usual methods of preserving or storing vegetables are given in the chart below.

VEGETABLE	METHOD OF STORING OR PRESERVING
Artichoke	Freeze
Artichoke, Jerusalem	Leave in the ground or freeze as a soup or purée
Asparagus	Freeze
Beans, broad	Freeze
Beans, bush	Dry, freeze, or pickle
Beans, green	Dry, freeze, or pickle
Beans, pole	Dry, freeze, or pickle
Beets	In boxes of sand or peat, or pickled
Broccoli	Freeze
Brussels sprouts	Freeze
Cabbage, red	Pickle
Cabbage, white	Freeze or store in trenches. (Savoys can remain in the ground until wanted)
Calabrese	Freeze
Carrot	In boxes of sand or freeze
Cauliflower	Will keep for three weeks if hung upside down in a cool spot, or freeze
Celeriac	In boxes of damp sand
Celery	Freeze
Corn	Freeze
Cucumber	Brine, pickle, or freeze as a soup
Eggplant	Will keep for two weeks in a cool place, or freeze immediately
Endive	Freeze in prepared dishes
Garlic	In slatted wooden boxes or tied in strings
Kale	Freeze
Kohlrabi	In boxes of sand
Leek	Leave in the ground or freeze
Okra	Freeze
Onion	Hang in strings, or in slatted wooden boxes
Parsnip	In boxes of sand
Peas	Freeze or dry
Pepper	Freeze
Potato	In baskets, or in burlap or paper sacks
Pumpkin	On shelves or hang in nets in frost-free place
Radish (winter only)	Boxes of sand
Rutabaga	Stacked in a cool, frost-free shed
Salsify	Freeze
Salsify, black	In boxes of sand
Shallot	Hang in strings or nets, or pickle
Spinach (all kinds)	Freeze
Summer squash	Hang in nets in a cool, airy place or freeze
Swiss chard	Freeze
Tomato	If green, in dark drawers until ripe, or freeze
Turnip	In boxes of sand
Zucchini	Freeze

A cool shed or basement is an excellent place to store fresh produce in sacks, boxes of sand, strings, or baskets. Make it as rodent-proof as possible. If the building has power, install a large chest freezer, too, so that you can store all your produce in one place. Always keep records to ensure that you use the oldest produce first.

Preserving Herbs

A few evergreen herbs, such as thyme and bay, can be picked and used year-round, but to enjoy most other herbs throughout the year, you will need to preserve them in some way.

The traditional method of doing this is to dry them. Some herbs can also be frozen and this is usually a more successful option for basil, chervil, parsley, and chives, which do not dry well. The flavor of stored herbs diminishes after a few months, so only keep small amounts.

Freezing herbs

A simple way to freeze herbs is to chop them and put them into ice-cube trays. Top off with water and freeze. They can be removed from the trays once frozen, and packed in freezer bags or containers. Use the frozen blocks to flavor soups, sauces, and stews; just stir them into the mixture as it is cooking.

Another way to freeze herbs (suitable also for basil, tarragon, and mint) is merely to wash and drain them dry before popping the sprigs into freezer bags. They will become limp as they thaw so are not suitable for garnishes. Instead, crumble them into the dish they are to flavor while they are still frozen. Herbs frozen in this way will keep only for about three months in the freezer before becoming discolored.

Drying herbs

If you are planning to dry herbs, be careful about when you pick them. All are best picked on a warm, sunny day in the morning after the dew has lifted. Pick them just before the flowering season; after this the leaves—which contain the flavor—toughen.

Choose only perfect specimens, with undamaged leaves. Discard any leaves that are withered or dead. You can dry herbs in a warm oven, spread out on wire racks or trays. Turn off the oven after an hour, but leave the herbs inside until it is cool.

Store dried herbs in a cool place, out of direct sunlight, and use as required. Remember their flavor is more concentrated than that of fresh herbs as the water content has been evaporated, just leaving their essential oils. When substituting dried herbs for fresh in a recipe, use a third or half as much.

Dry sprigs of herbs in small bunches to see you through winter. Only dry a little of each, as they will not keep for more than a few months.

AIR-DRYING HERBS

1 Tie herbs in small bunches. Dip them for a few seconds in boiling water and then refresh them quickly under the cold tap. This helps to clean them and to retain their color, but it is not essential. Shake off excess water and dab them on paper towel to get them as dry as possible.

2 Hang the bunches in a warm, dry place out of direct sunlight for about three days. Or spread them on a wire rack and leave in a warm cupboard for up to five days.

3 When the herbs are quite crisp and brittle to the touch, they are dry. Either crumble them with your fingers or crush them with a rolling pin, discarding any tough stalks. If you want a finer powder, pass them through a sieve. Store the crumbled herbs in air-tight jars and keep them in a cool, dark place.

Preserving Fruit

The only fruits that can be kept raw for longer than a month are the later varieties of apple. All other fruits must be preserved if you cannot eat them fresh from the tree or bush.

The most common ways to preserve fruit are freezing and canning. Pickling, jam, and jelly-making are other methods useful for using up a glut of fruit. A few fruits may also be dried.

DRYING

This method is particularly suitable for apricots, peaches, and plums, but apples, pears, and grapes may be dried too.

The principle of drying fruits is much the same as that for drying vegetables (*see* page 208), although if you use an oven rather than a dehydrator, do not let the temperature of the oven rise above 120°F (50°C) for at least the first hour, otherwise the skins will either harden or burst apart.

You can also dry fruits at normal air temperature. Prepare the fruits as you would for dehydrating for oven drying then spread them on wire racks or thread onto lengths of dowel and leave in a warm, dry place until shriveled and dry.

Using dried fruits

You can use dried fruits just as they are or snack on them like chips, but for cooking they are often brought back to a more moist state. Soak the fruit in water, with sugar added if you like, until the pieces are soft and ready to use.

BEST METHODS FOR STORING FRUIT

The most successful and usual method of preserving or storing fruit is given in the chart below.

FRUIT	METHOD OF STORING OR PRESERVING
Apple	Store later varieties wrapped in waxed paper in boxes; otherwise freeze or dry
Apricot	Can be kept for up to a month in wooden trays in a cool, airy place, providing they were only just ripe when picked. Otherwise, freeze, can, pickle, or make jam
Blackberry	Freeze, bottle, can, or make jam
Black currant	Freeze, can, or make jam or jelly
Blueberry	Freeze
Crabapple	Pickle, or make jam or jelly
Cherry	Freeze, can, or make jam
Red and white currant	Freeze, can, or make jelly
Fig	Freeze, can, or make jam
Gooseberry	Freeze, can, or make jam
Grape	Dry, or make jelly or wine (*see* page 230)
Lemon	Can or make jam
Loganberry	Freeze, can, or make jam
Melon	Freeze
Peach and nectarine	Freeze, can, or make jam
Pear	Freeze, can, dry, or make jam
Plum	Freeze, can, dry, or make jam
Quince	Make jelly
Raspberry	Freeze, can, or make jam
Rhubarb	Freeze or can
Strawberry	Freeze, can, or make jam

Apricots, peaches, and plums

Choose the largest varieties and unblemished fruit. Wash the fruit if necessary, then halve it and remove the pits. If you do not have a dehydrator, cover wire racks with cheesecloth or foil and lay the fruit out on these, in single layers, not touching. Place in an oven heated to the temperature recommended opposite and leave until the skins start to shrivel then raise the temperature very slightly.

Plums can be frozen, canned, dried, or used to make jam. Cut them in half and remove the pit before preserving them.

Thread apple rings onto a length of wooden dowel and hang it somewhere warm and dry until the fruit is ready.

OVEN-DRYING FRUIT

Commercially dried fruit is often prepared using varieties specially developed to retain their color when dried. Don't despair if your own efforts look less than perfect—they will still taste delicious. You can follow these steps for drying root and tuber vegetables, too. Blanch vegetables, except for tomatoes, sweet peppers, okra, mushrooms, beets, and onions, before drying. For mushrooms, remove the stalks.

1 Start with only good quality, firm, blemish-free fruit that is just ripe. Peel, core, and slice as necessary, cutting very thin and even slices.

2 For fruits that discolor, such as apples and pears, dip the slices in a weak solution of lemon juice: 6 tablespoons of lemon juice to 1 pint (500 ml) of water.

3 Lie the slices on cheesecloth-covered racks over a foil-lined baking sheet, making sure that they do not touch. Arrange halved fruits cut-side down. Put the oven on its lowest setting, put the fruit in, and leave the door slightly ajar.

4 Dry for the times given in the table (right), turning midway through and switching the position of trays in the oven. The food is ready when it is dry and leathery.

5 Leave to cool completely before storing. Pack slices in airtight containers, layered between parchment paper, and store in a cool cupboard.

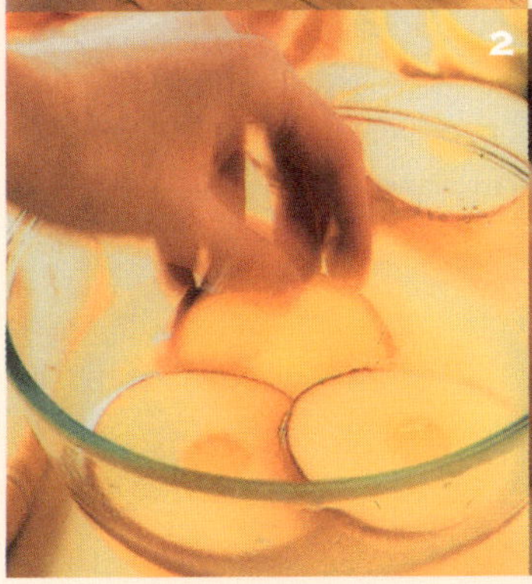

OVEN-DRYING TIMES FOR FRUIT

With your oven on its lowest setting, dry sliced and prepared fruits (and vegetables) as follows

FRUIT	TIME
Apple rings	6–8 hours
Apricots, halved and pitted	36–48 hours
Bananas, peeled and halved lengthwise	10–16 hours
Berries, left whole	12–18 hours
Cherries, pitted	18–24 hours
Herbs, tied in bundles	12–16 hours
Peaches, peeled, halved, and pitted	36–48 hours
Peaches, sliced	12–16 hours
Pears, peeled, halved, and cored	36–48 hours
Pineapple, core and cut to ¼ in (5 mm) rings	36–48 hours
Plums, halved	18–24 hours
Vegetables, ¼ in (5 mm) slices	2 hours
Vegetables, ½ in (10 mm) slices	7–8 hours

These fruits are dry when no moisture comes out of them and the skin does not break when you squeeze them gently.

Apples

The best way to dry these is in rings. Choose crisp, sweet apples; peel and core them, then slice them into ¼ in (5 mm) rings. As you cut these, put them into some lightly salted water or a lemon juice solution to help prevent discoloration. Pat the apple rings dry on sheets of paper towel, then thread them onto thin wooden rods or dowels. Hang these in the oven heated to the temperature given on page 221. They should dry in about six hours, by which time they will feel leathery and look dry and shriveled on the outside. Spread them out on a wire tray and leave for a good 12 hours to dry.

Pears

Choose ripe, but firm fruit and peel and quarter them or cut into slices. Cut out the cores and drop the quarters into lightly salted water or lemon juice to help prevent discoloration. Spread out on cheesecloth-covered wire trays and put these in the oven heated to its lowest setting. They will take about the same amount of time to dry as the apple rings.

Grapes

The seedless varieties are the best for drying; they should be just ripe. Wash them and pat them dry on sheets of paper towel. Do not cut the fruits. Spread out individually on cheesecloth-covered wire trays and put in an oven heated to its lowest temperature setting. The grapes are dried when the skins are shriveled and do not burst when you squeeze them gently.

FREEZING

Nearly all fruit freezes well, although different methods suit different types of fruit. For all methods, choose ripe but not overripe fruits that are not bruised or blemished in any way. Fruit is best frozen within a couple of hours of being picked.

Think about how the frozen fruit is likely to be used, too. If it is to be eaten whole in its straight, unfrozen state, it is best to open-freeze it, although this is the most time-consuming method. Fruit that is to be used in cooked pies may be sugar-frozen, and if it is intended for mousses and similar dishes it is best to freeze it puréed.

Most frozen fruit must be defrosted before using it in any way, although soft fruits that are to be used raw in a dessert can be thawed to a chilled state. Fruit that is to be put into smoothies can be used frozen. If you are going to stew fruit, it can be gently heated from frozen.

Open-freezing

This is most suitable for soft fruits, such as strawberries, raspberries, and blackberries. As with vegetables that are open-frozen, each fruit remains separate, so that you can take only as much as you need at one time from the freezer bag. Wash the fruit only if absolutely necessary (the drier the fruit, the better the frozen results), remove the stalks, then place each piece of fruit separately on a tray; they should not touch one another. Freeze until the fruit is firm, pack it in freezer bags, seal it, and then label it before returning it to the freezer. When thawed, the fruits will be considerably softer than if fresh, but will still retain their shape.

Golden rules for freezing

If you are freezing more than 2 lbs (1 kg) of fruit at once, turn your freezer thermostat to "fast freeze." Always make sure that food is completely cold before putting it into the freezer. Pack food tightly into rigid containers, trying not to leave any gaps. If there is room left in the container and you have no more food to freeze, fill the space with crumpled paper. Remember to leave a 1-inch (2 cm) expansion gap for liquids. Sauces and purées can be stored in plastic bags, but to avoid awkward shapes, put the bags into boxes before filling and freezing them, then lift them out and seal once the contents are solid.

Sugar-freezing

This method is also suitable for soft fruits, but the fruit will be very mushy when thawed and is best used in cooked desserts. Wash the fruit if necessary, remove the stalks, then either roll each fruit in superfine sugar, or put them in a dish, sprinkle with sugar (add as much as you like according to taste), and stir gently when the sugar has dissolved.

Pack the fruit into rigid containers, seal, label, and freeze. Another way of sugar-freezing is to layer fruit and sugar in alternating layers in rigid freezerproof containers.

Freezing in syrup

This is suitable for pitted fruits and those that tend to be less juicy, such as apples and pears. They can be frozen in a heavy or light syrup, according to taste, what you want to use the fruit for, and the type of fruit. Soft fruits are best frozen in a heavy syrup, while the more delicately flavored fruit, such as pears or melons, are better in a light syrup.

For heavy syrups, dissolve 1 lb (450 g) of sugar in each 1 pint (575 ml) of water; for a lighter syrup, halve the amount of sugar. When the syrup has completely cooled, pour it over the prepared fruit, packed into a rigid container. Make sure the fruit is submerged in the liquid before putting on the lid by placing a piece of crumpled wax paper on top of it.

It is a good idea to poach most pitted fruits—apricots, peaches, and plums—before freezing, or else the skins are likely to become tough during the freezing process. They need only a few minutes simmering in a heavy syrup. When they have cooled completely in the syrup, they can be frozen in the normal way.

Freezing in a purée

Puréed fruit is useful for all sorts of cold mousses, soufflés, ice cream, gelato, and sauces. It is also a good way to freeze fruit that is slightly overripe or imperfect (although all bruised or damaged parts should be discarded).

Soft fruits, such as raspberries, strawberries, and blackberries, can be pushed through a sieve and frozen immediately. Sugar can be added before freezing, or after thawing and before use.

FREEZING FRUIT

FRUIT	METHOD OF FREEZING
Apple	Sugar-freeze; poached in light syrup, or as a purée. If sugar-freezing or freezing in a light syrup, apple slices can be prevented from discoloring by dropping them into a bowl of water and lemon juice and adding lemon juice to the syrup.
Apricot	Sugar-freeze; poached in heavy syrup, or as a purée.
Blackberry	Open-freeze; sugar-freeze, or as a purée. They can also be frozen in a heavy syrup if wanted for pies or crumbles.
Blueberry	Open-freeze; sugar-freeze, or in a light syrup.
Black currant	Open-freeze; sugar-freeze, or as a purée.
Cherry	Open-freeze; sugar-freeze, or in a light syrup. Choose red varieties and pit them before freezing.
Fig	Open-freeze, or freeze in a light syrup.
Gooseberry	All methods are suitable; choose according to the use for which they are required.
Melon	Cut flesh into cubes or balls and cover in a light syrup.
Loganberry	All methods are suitable; choose according to the use for which they are required.
Lemon and other citrus fruits	Citrus do not generally freeze well. Squeeze the fruits, collect the juice, and freeze the juice in bottles or icecube trays.
Peach and nectarine	Sugar-freeze, or poached in heavy syrup.
Pear	In light syrup (treat as for apples to prevent discoloration).
Plum	Sugar-freeze; poached in heavy syrup, or as a purée.
Raspberry	All methods are suitable; choose according to the use for which they are required.
Rhubarb	Open-freeze (but the cut pieces should be blanched for one to two minutes first); in heavy syrup, or as a purée.
Strawberry	All methods are suitable; choose according to the use for which they are required.

Other, more robust fruits—gooseberries, black currants, apples, apricots, and others—must be lightly cooked before they can be liquidized or sieved, or the results will not be smooth. The purée must be quite cold before freezing.

To make the most efficient use of your freezer space and your plastic containers, freeze sauces and purées in bags contained within boxes. Pour in the liquid, freeze it, and then lift out the bag with its frozen "brick," seal it, and stack it with your other frozen produce.

Canning Fruit

Most fruit can be canned, and it makes an invaluable standby for winter desserts. The bottles or jars should be made of thick glass with either a screwtop or a secure, clip-on lid.

The USDA has extensive canning guidelines available, especially for beginners who have little experience preserving fruit. It's worth checking out their website for downloadable instructions. Canning low-acid fruits can be potentially harmful if the correct process is not followed. If you have never canned fruit before, be sure to check (and double-check) the recipe and additional guidelines to be sure you are processing for the correct amount of time and with the right equipment.

First, check your jars

When planning to can fruit, check the jars carefully first to ensure none have chips around the ring. If they do, the seal will not be airtight. Make sure the lids are in good condition too—that glass is not chipped or cracked, metal is not bent or rusty, and that any spring clips are strong.

Test them before use to ensure that the seal is airtight by filling the jars with water, putting on the lid, and securing it with the screw ring or spring clip. Then stand the jars upside down on a table. If there is any sign of leaking after five minutes or so, the jars should not be used.

Sterilize jars before using with boiling water just before filling them with the fruit, but do not dry the inside, or you will spread germs, even with the cleanest dish towel. Instead leave them to drain turned upside down.

CANNING FRUIT

1 Prepare the fruit as required (*see* below) peeling, coring, or cutting as necessary. Blanch peaches and apricots first to make it easier to remove their skins.

2 Pack the fruit evenly into the jars, using the handle of a wooden spoon to push down and position the firmer fruits, such as apricots, if necessary.

3 Fill the jars with the firm fruits and then pour in the syrup, jerk the bottles sharply to release any air bubbles and top off with more syrup. When bottling soft fruits, it is better to fill the jars a quarter at a time, alternating fruit and syrup until the jar is full. This gives a better distribution of fruit and syrup.

4 Tighten the lids, and if they are screw tops, undo them a quarter turn—this prevents the bottles bursting during the sterilization process that follows (spring clips are designed to avoid any danger of this happening).

Choosing the fruit

Use only perfect, unblemished fruit and grade it according to size so that fruit of a similar size is canned together. Prepare it according to its type; apples and pears should be peeled, cored, and sliced or quartered (put them into cold, lightly salted water or lemon juice solution as you prepare them to prevent discoloration); apricots and peaches should be skinned, halved, and pitted; cherries and plums can be canned whole, or plums may be halved and pitted if you prefer—wash them all first.

Rhubarb should be cut into small chunks and will pack better into the jars if it is steeped in hot sugar syrup overnight. All soft fruit such as black currants, raspberries, loganberries, and strawberries should be canned whole, with the minimum of handling. Some people say that the flavor of canned strawberries is better if they are soaked overnight in syrup and their color can be improved by adding a few drops of edible red coloring to the syrup.

Making a sugar syrup

Fruit is usually bottled in a sugar syrup, and although there is no reason why it cannot be bottled in plain water, the flavor will not be as good. Make a syrup with about

PROCESSING TIMES FOR CANNING FRUIT

HOT WATER BATH METHOD
2 minutes Apple slices, blackberries, black currants, gooseberries, loganberries, raspberries, rhubarb, strawberries (unsoaked)
10 minutes Apricots, cherries, whole plums
20 minutes Peaches, halved plums, strawberries (soaked)
40 minutes Pears

OVEN METHOD
30–40 minutes Apple slices, blackberries, black currants, gooseberries, rhubarb
40–50 minutes Apricots (whole), cherries, loganberries, raspberries, strawberries (unsoaked)
50–60 minutes Apricots (halved), plums (whole), strawberries (soaked)
60–70 minutes Peaches, pears, plums (halved)

8 oz (225 g) of granulated sugar to 1 pint (575 ml) of water—vary the strength according to the fruit and your taste. You can add flavorings to the syrup if you wish, such as vanilla pods, or spices, such as cinnamon.

STERILIZING BOTTLES

Bottled fruit must be sterilized to prevent the growth of damaging organisms and bacteria.

There are many ways of sterilizing bottles, but two of the simplest are described here.

Hot water bath, quick method

Pack the fruit into jars and fill them with hot syrup before securing the tops. Put them into a deep pan or any other large pan in which there is a false bottom (a fish poacher is ideal if you do not have a special sterilizing pan); if the pan does not have a false bottom, you can use canning racks or improvise with thickly folded newspaper, cardboard, or a towel in the bottom.

The bottles must not touch one another—separate them with pads of folded newspaper. Pour warm water (just above body temperature) into the pan to cover the bottles and put the lid on. Bring it up to simmering point—190°F (88°C)—for about 30 minutes. Remove the bottles using bottling tongs (or tip some water out of the pan and hold the jars with an oven mitt) and secure the lids. Leave the bottles overnight to cool, then test the seal.

Oven method, wet pack

Pack the fruit into jars and top off with boiling syrup to about ½ in (1 cm) from the tops. Dip the glass lids and the rubber bands in boiling water and put them on the jars, but do not screw on the metal tops or fasten the clips. Stand the jars on a piece of cardboard or some thickly folded newspaper in the center of an oven heated to 300°F (150°C). Make sure there is at least 2 in (5 cm) between the bottles at all points, then leave them for the times given on the left. Remove them one by one, preferably onto a wooden surface and either fasten the clips or screw on the metal tops. Test the seal the next day.

Testing the seal

Between 12 and 24 hours after processing the fruit, you can test the seal. Unscrew the metal top, or release the clip, and pick up the bottle by the glass top only. (Be ready to catch the bottle). If the top remains in place, the seal is perfect. Label the bottles and store them in a cool, fairly dark place. If the lid comes off, the seal is not good. The fruit can be reprocessed, but it is better used immediately.

Bottling fruit in alcohol

Prepare the fruit and pack it into clean, sterilized jars, layering it with sugar as you go, then fill the jars with alcohol, shaking to remove any air bubbles and topping off again once more. You can use any alcohol that is more than 40 percent proof, either using it on its own or combined with a sugar syrup to make a richer liquid.

Seal the jars and store them somewhere cool, dark, and dry for at least two months—but the longer, the better. To help the sugar to dissolve, shake the jars every few days for the first month.

Cherries in brandy make a traditional and always well-received gift. Choose your own combinations of fruit and liquor, adding other flavors if you like.

Making Jams and Jellies

This is another way of preserving fruit, although it changes the nature of the fruit even more than canning does. Jam is a popular way of using up cultivated and wild-gathered fruit.

Jam is made by boiling fruit and water together until it thickens when it cools into a jellylike preserve. This is called the setting point. The set comes from pectin—a natural substance contained in fruit, which once released, reacts with the sugar.

Some fruits are rich in pectin, such as apples, currants, cranberries, and citrus. Others are poor, such as cherries, rhubarb, pears, and strawberries. In between these are a number of fruits—apricots, blackberries, plums, loganberries, and raspberries—that have a medium pectin content. This means that unless additional pectin is added to the fruit, the jam will be less firm than that made with pectin-rich fruits.

As with canning, always follow safety guidelines for making fruit to avoid any spoiling. All equipment should be sterilized and seals on jars must be airtight.

Choosing your fruit

Even though fruit for jam-making is to be cooked to a pulp, only good-quality, unblemished, fresh, just-ripe fruit should be used. Overripe fruit tends to lose its pectin, and damaged or bruised fruit could affect the taste of the jam. De-stem, wash, and drain soft fruit such as raspberries and strawberries. For harder fruit like apples and pears, peel, core, and slice the fruits. For softer pitted fruit like apricots and plums, cut them in half, remove the pits, and cut the flesh into quarters. Try to cut all fruit into similar-size pieces for even processing.

BASIC TECHNIQUES FOR JAM-MAKING

Whatever your choice of fruits, the method of jam-making is the same. Experiment with combinations of flavors—mixing berries with ginger, for instance, or adding some spices to give the jam a bit of heat.

MAKING JAM

1 Sort through your fruits and discard any that are overripe, damaged, or blemished. Prepare as necessary, by chopping, peeling, coring, or pitting.

2 Put the fruit in your jam pan with a little water if necessary and start to heat it. Remove from the heat to add the sugar, then return to the stove and heat it gently, stirring, until the sugar has dissolved.

3 Bring to a boil and boil rapidly, without stirring, until the jam reaches setting point (*see* page 228). This usually takes between 10 and 15 minutes, depending on the quantity you are making.

4 Let the jam cool a little, remove any scum from the surface, and leave the jam to sit for another 10 minutes until it is cool enough to handle safely.

5 Spoon the jam into warm, sterilized jars (a funnel and ladle make this job much easier and less likely to burn your hands), top with wax paper, and seal and label the jars.

When you make your own jams and jellies, you can try delicious combinations that you cannot get in the supermarket, such as pear and raspberry.

The fruit should be cooked in a large, uncovered pan (*see* page 212) over a gentle heat, adding water if necessary to stop the fruit sticking. Most soft fruits, such as raspberries and strawberries, do not need any additional water, but the individual recipes will indicate this.

Add sugar according to the recipe after the fruit has cooked down into a pulp and reduced in volume. Use granulated or preserving sugar, which dissolves quickly, making sure it is thoroughly dissolved. This will vary according to the pectin content of the fruit, so be sure to follow the recipe exactly, rather than adding more or less to suit your taste.

If you need to add pectin, you can buy it in liquid or powdered form in most large supermarkets or at a hardware store or farm-supply store. There are different types, including low-sugar and sugar-free pectin; experiment to find which give you the best results. You can also buy special jam sugar, although ordinary granulated sugar is fine to use.

Finishing and canning

When it reaches the setting point, let the jam stand for a few minutes and then skim the scum off the surface, using a slotted metal spoon. A pat of butter stirred into the jam will remove all final traces of scum, and the jam can then be poured into warm, clean jars.

Fill the jars to within ¼ in (0.5 cm) of the rim and seal the jars. Process the jam to prevent spoilage in a boiling water canner for five minutes or as directed in the recipe. (If you are at a higher alititude, add to the processing time.) If you are planning to eat the jam right away, you may not need to process the cans, but seal the jars and keep the jam in the fridge.

Testing for the setting point

The jam should be kept boiling until it reaches the setting point. This can be determined by temperature if you have a special jam or sugar thermometer: the setting point is 220ºF (104ºC). Alternatively, use one of the following simple tests.

Flake test

Take out a spoonful of jam, using a wooden spoon. Hold it for a minute to cool it, then turn the wooden spoon sideways; if the jam drops off the edge of the spoon in large flakes, the setting point has been reached.

Saucer test

Put a little of the jam onto a cold saucer and let it cool. If the surface of the jam forms a skin that wrinkles as you push it gently with your fingers, the setting point has been reached.

Keep a saucer in the fridge while you are cooking your jam, ready to test for the jam's setting point. If a skin wrinkles on the jam when you push it gently, it is ready.

Jam can be ladled straight into wide-necked jars, but remember that it will be very hot and if you spill any while holding the jar steady, you will burn your hand. A jam funnel, with a wide neck to let pieces of fruit through, will save both time and accidents.

MAKING JELLIES

Fruit jelly often uses the same basic ingredients as jam, but the result is a smooth, clear preserve. However, mkaing jelly does require more fruit.

To achieve this, the fruit is placed into a jelly bag after it has been cooked, and allowed to drip slowly into a bowl overnight, or for as long as 24 hours.

It is important to let this process happen slowly. If you are impatient and squeeze or prod the bag too vigorously, particles of pulp will also come through, and the jelly will be cloudy. You can buy jelly-bags with their own stands through mail-order kitchen suppliers.

The liquid is then boiled with a precise amount of sugar, according to the volume of the juice, until the setting point is reached. As only the liquid from the fruit is used, similar quantities of fruit make considerably less jelly than jam.

MAKING MARMALADE

If you have citrus trees that produce enough fruit, make your own estate marmalade, but supermarket fruit will work just fine.

Even if you cannot produce your own fruit, once you become proficient at making jams and jellies, you may like to try something different and make a batch of marmalade.

The main difference between making jam and marmalade is that marmalade incorporates pieces of chopped peel, and this must be cooked separately first to release the pectin and tenderize the skin.

MAKING MARMALADE

1 Cut the fruits in half and squeeze out the juice, saving any pits and pulp and wrapping it in a piece of cheesecloth or a muslin bag.

2 Slice the peel according to taste and put it into a pan with the juice and water, as specified in your recipe. Suspend the bag of pulp in the water, too. Bring to a boil then simmer for one and a half hours, until the peel is soft.

3 Squeeze the liquid from the bag of pulp and discard it, then add sugar to the pan. Heat gently, stirring to dissolve the sugar, then boil hard for up to 15 minutes, until the setting point is reached.

4 Skim off any scum on the surface and put the marmalade in jars. Process as needed.

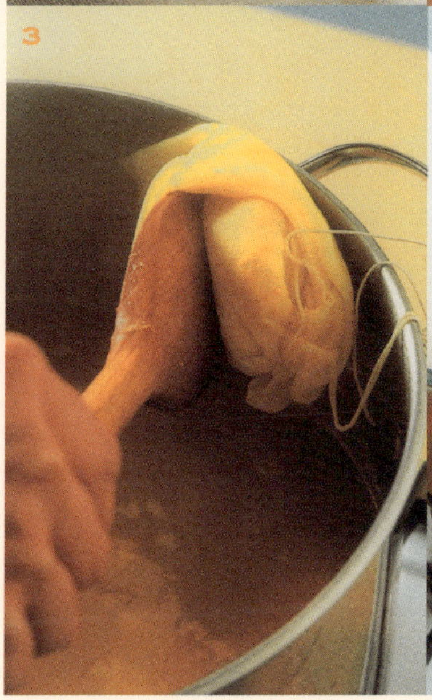

Making Wine and Cider

Making your own wine and cider is an excellent way of using some of the produce you have grown or collected. You don't need a grapevine, but can make wine from a wide variety of ingredients to suit your own tastes.

Berries, fruit, flowers, vegetables, and herbs can all be used to make wine, although some flowers and berries must be avoided as they are toxic, and not all vegetables make very palatable wine. Flowers and fruit that are worth experimenting with include rose petals, elderflower, dandelion, cranberry, huckleberry, kiwi, primrose, rhubarb, persimmon, and many more.

The main vegetables to avoid are pumpkin, squash, potato, tomatoes, and turnips, although all vegetable wines tend to be an acquired taste.

Glass carboys and airlocks are essential pieces of equipment for making wine and you will probably need several of each. You can reuse old wine bottles or buy them at wine-making suppliers, together with corks.

Find your own blends

Amateur home winemaking is an experimental hobby: tastes are very individual and weights and measures are rarely critical. The steps to follow are the same whatever the ingredient you use and, as you get more confident you will find you need to turn to the recipe books less and less, following your own instincts and being inspired by the ingredients you have on hand.

It is always a good idea to make small quantities of wine each time—about 1 gallon (4.5 liters) is a sensible amount. If it proves undrinkable, it is not a terrible waste when you have to throw it away.

Cleanliness and patience are key

The two most important aspects of winemaking are absolute cleanliness and considerable patience. All the equipment (*see* below) should be sterilized before use, otherwise you run the risk of attracting insects or allowing bacteria to grow in the developing wine. Both will give the wine a strong vinegary taste, rendering it quite useless for drinking.

Sterilizing is an easy process and all home winemaking suppliers sell various sterilizing solutions that have clear instructions for use. It is a good idea to sterilize equipment when you have finished using it, before putting it away, and then again when you are ready to use it the next time. After sterilization, rinse the equipment in cold water and leave it to drain, rather than drying it on a cloth.

Patience is extremely important. If you do not leave your wine for long enough to mature (at least six months before bottling) it will not be worth all the effort. In many instances it improves still further if left for longer. If you make small quantities of wine frequently, you will soon find that you always have some fully-aged wine ready to drink, so you do not need to be impatient with the more recently-made vintage.

BASIC INGREDIENTS

As well as the main fruit, vegetable, flower, berry, or herb in your wine, you will need the following ingredients. Anything not readily available from your cupboards or the supermarket can be purchased online, at a grocery store, or through a home-winemaking supplier.

Water According to each individual recipe. Some ingredients will need more additional water than others.

Sugar It is this which determines the alcohol content of the wine as well as the sweet or dry taste. Sweet or dry wine can be made from any ingredients, the deciding factor being the amount of sugar added.

Many recipes, old-fashioned ones in particular, tend to produce wine that is sweet, so if your taste is for a drier wine, you may find you need to reduce the amount of sugar considerably. As a guide, 2½ lbs of sugar per gallon (250 g of sugar per liter of liquid) produces a dry wine; 3 lbs of sugar per gallon (300 g of sugar per liter) produces a medium wine; and 3½ lbs of sugar per gallon (350 g of sugar per liter) produces a sweet wine, although this should be slightly adjusted according to the natural sweetness of the chief ingredient. You can use ordinary granulated sugar.

Yeast This brings about the fermentation process that turns the liquid into wine. Various types of yeast are available, although some will produce better results than others. Ordinary brewers' yeast can be used, but it is better to use a special wine yeast; these are readily available through home winemaking suppliers (*see* below).

In addition, there are numbers of particular yeasts produced in the various winegrowing districts of the world. Experiment with the various kinds to find those you like best.

Nutrient Yeast is a living organism and needs feeding if it is to grow and do its work. The nutrient it needs to do this job of fermenting the wine is commercially available and the quantity to be added to the wine will be listed on the package.

Acid The yeast needs acid conditions to bring about good fermentation. Most homemade wines need acid added to them, usually in the form of citric acid. Follow individual recipes for how much to add—it will depend on the natural acidity of the ingredients. Some need only the juice of one or two lemons.

Tannin This is a substance found in the skins of fruit, particularly those of red fruit. It gives wine its characteristic "bite." Wines made from red fruit and berries—such as elderberries and plums—will not need extra tannin; most other wines, even white wines, will benefit from the addition of grape tannin, which you can buy from wine-making suppliers, although if you prefer, add strong, cold tea instead.

Pectin-destroying enzymes These should be added to wines made from fruit that is particularly high in pectin (such as apples, black and red currants, and gooseberries), which can make the wine cloudy.

Sterilizing tablets These can be used for sterilizing equipment and they are also added to the wine (*see* below) to kill off any lingering bacteria. Choices include sodium metabisulfite, potassium metabisulfite, and Campden tablets.

ESSENTIAL EQUIPMENT

Making wine requires some special equipment and it is very difficult to make decent wine without using proper winemaking supplies. Instead of trying to improvise and compromise, you can often find much of what you need for sale second hand.

It is a good idea to keep things simple when you first start, in case you don't wish to continue beyond the first sip of your first bottle. Even so, you will need to purchase some key items. If you enjoy winemaking, you can always buy more specialized equipment as you go on.

At the very least, you will need some fermenting buckets, preferably with spigots, glass carboys and bungs, a hydrometer (for testing the sugar levels and, therefore, the strength of your wine), plenty of empty wine bottles and new corks, a funnel, airlocks, siphon tube, filters, and a corker and capper.

HOW TO MAKE WINE

The basic procedure for winemaking is the same whatever kind of wine you are producing. Always choose top-quality, unblemished ingredients that are fresh and ripe.

Start by boiling or steeping your vegetables or fruits (right); flowers should be stripped from the plant, placed in a container, and bruised with a wooden spoon before boiling water is added to them. They, too, are left to steep for a few days.

Some recipes stipulate adding sugar and a sterilizing tablet at this stage (with apples, this tablet helps to prevent the fruit discoloring or beginning to oxidize). This is to extract the maximum amount of flavor from the ingredients, but stirring the mixture each day will also help. Be guided by common sense: strawberries turn mushy very quickly and release all their flavor to the liquid within a couple of days; harder fruit or berries will need longer.

Preparing the yeast

Get the yeast working in a starter bottle before adding it to the wine. Put the yeast into a small bottle with some fruit juice, sugar, and nutrient, plug the top with a piece of cotton wool, and leave it in a warm place. Some types of yeast will begin to activate within a few hours; most take a couple of days.

Racking and bottling the wine

Ferment the wine in a glass carboy (right) until the airlock has stopped bubbling. Shake the jar, then leave it for another couple of days—but no longer. The next stage is to filter the liquid into a clean, sterilized jar, leaving the sediment at the bottom of the old jar. This process is known

MAKING WINE

1 Wash and peel the vegetables or fruit you are using for the wine, and then chop into small pieces. Simmer vegetables in water until tender or steep fruit in a large container of boiling water, cover, and leave until the fruit is thoroughly mushy.

2 Yeast can be added at this stage of the process. It is mixed with fruit juice, sugar, and nutrient in a small bottle of its own and added to the wine after a few days.

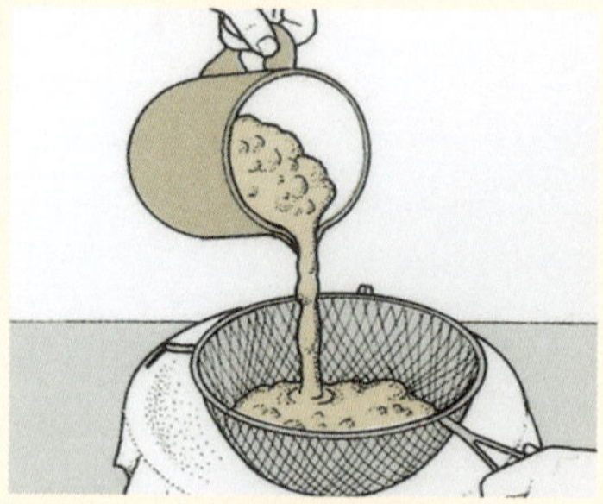

3 When the ingredients have steeped, strain the liquid through a sieve and a filter. Add the yeast, sugar, nutrient, and other ingredients if you have not done so already.

4 Strain this liquid, known as the "must," through a piece of filter in a plastic funnel into a sterilized fermentation jar or carboy.

5 Carefully fit an airlock into the top of the carboy. Half-fill it with distilled water, to which is added a quarter of a sterilizing tablet. Keep the wine at around 70°F (21°C)—too hot and the wine will spoil, too cold and fermentation will slow down or stop.

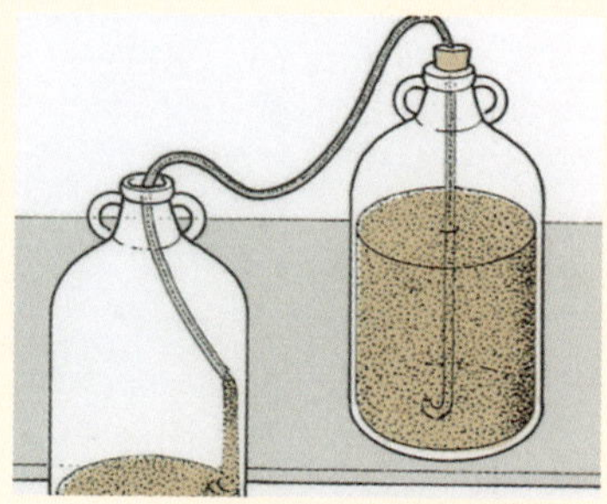

6 When fermentation has ceased (the airlock stops bubbling), rack the wine by siphoning the liquid into a clean carboy and seal it firmly with a bung. Do this at least twice more to remove all sediment before finally siphoning it into bottles, corking, and labeling.

as racking. Place the clean jar at a lower level and siphon off the wine from the old jar through a tube.

Taste the wine at this stage, for if it is too sharp or dry, you could add some concentrated syrup or sugar dissolved in water. Regardless, top off the jar with water and add a sterilizing tablet to ensure that no bacteria is allowed to get to work. This jar should then be fitted with an airtight rubber bung and stored preferably in a place where the temperature is about 70°F (21°C).

The wine should be left for at least six months before it is bottled. If, during this time, a heavy sediment forms at the bottom of the jar, it should be racked again. It does not matter how many times you rack the wine—and it is a good idea to do so at least two or three times. The more you do, the better the wine is likely to be, as it helps to reduce any risk of the sediment at the bottom of the jar tainting the wine.

Bottling

The final process is to pour the wine into sterilized bottles. The corks, too, should be soaked in a sterilizing liquid. If you soak them for 24 hours in cold, boiled water, they will also soften, which makes them easier to drive into the bottles.

The wine should be filtered and then poured into the bottles so that it is about ¾ in (2 cm) below the cork. Drive the cork into the bottle, then label the wine with the name and date of bottling. Wine should always be stored on its side, so the cork is kept moist. If it is allowed to dry out, bacteria is able to enter the bottle and the wine will turn vinegary. No wine should be drunk for at least one month after bottling, and it is advisable to leave it much longer if you can.

MAKING CIDER

If you have an abundance of apples and have made enough apple wine, you can turn the remainder into cider. Pears can be made into perry.

In both cases, it is only the juice of the fruit that is used—no yeast, water, or sugar (unless you want to speed the fermentation or produce a very sweet cider). The best cider will be made from a mixture of sweet and more sour-tasting apples, and you will get the greatest amount of juice if you let them soften a little first. This does not mean leaving them to rot; if you have many heavily bruised apples, the cider will not have a good taste.

Crushing your apples is the first stage in making cider. You can do this by hand with a mallet or in a proper cider press.

The apples need crushing. If you find you enjoy your cider and the process of making it, and have a lot of apples to spare, it may be worth investing in a cider press; these are expensive pieces of equipment, but it is possible to make your own.

An alternative is to crush the apples with a wooden mallet, wrap the pulp in muslin or cheesecloth, and press it to extract the juice. Another method is to liquidize the apples or put them through a food-processor or blender then press them through cloth in the same way. You can even push the wrapped-up pulp through a mangle.

After that the juice can be poured into a clean, sterilized carboy or glass jar. Lay an inverted saucer over the top and leave the juice to ferment. If you want to speed the process, you can add yeast in the way you would in winemaking, but this is not absolutely necessary.

Rack the cider (*see* page 232) after fermentation has finished, and only bottle it after it stops giving off gas. It will improve if left to age for several months before drinking.

When bubbles have stopped escaping through the airlock, the cider is ready for bottling, but let it rest for a while before you drink it.

FLAVORED GIN AND OTHER ALCOHOL

Sloe gin is a traditional favorite, historically steeped through the winter to be ready in time for Christmas, but you can use other base spirits and ingredients to make your own liqueurs.

You don't need good quality alcohol to make sloe gin, but flavored vodkas will taste smoother if you start with a moderately good base spirit. Citrus vodkas are popular, but you can infuse either spirit with any flavors you like or have available, such as blackberries, orange, raspberry, apple, or liquorice.

Slice your fruits if they have skins, so that the flesh is exposed. Sloes should be pricked with a pin to break the skin, although you can also put the fruits in the freezer to split them.

For flavoring vodka, remove any rind, seeds, or pits, put the fruit in a large jar or other sealable container, and add the vodka until the fruit

Sloe gin is a potent and warming spirit, best taken in small measures, but also makes a fruity contribution to cocktails.

Homemade wines and spirits make great presents and are fun to share with guests. Try making something unique with your own particular blend of flavors.

is all submerged. Steep the vodka for about two weeks, shaking every couple of days, then strain through a jelly bag or coffee filter paper and bottle. Allow the vodka to rest for up to four weeks before drinking.

Sloe gin

This gin is more like a liqueur than a spirit, as it includes a quantity of sugar and is allowed to steep for several months. Prick the skins of the sloes and put them in a carboy or several large bottles until they are half full. Add the same weight of sugar and top off with the gin.

Shake the bottles every day for the first week and then every week for the next two or more months. The longer you leave it before straining and bottling, the better the flavor of the gin.

After you remove the sloes from the liquor, you can use the steeped fruits in pies or sauces, although they will have taken on some of the taste of the alcohol.

Goat Milk

Goat milk can be made into all the dairy products—cream, butter, yogurt, and various cheeses—most usually associated with cow milk, but cheese is usually the most successful.

Most people find it impractical to make cream and butter from goat milk because of the amount of milk needed (plus the quantity of whey left as residue), and the expensive equipment that must be used. One gallon (4½ liters), for example, makes only about ¾ to 1½ pints (450 to 900 ml) of cream, which is essential to make good quality butter. It is hardly worth doing so unless you can use 1 gallon (4.5 liters) of cream at a time and for this you would need at least four goats in prime milking condition—too many for the average household. Never mix the milk collected on different days, as this will result in a very "goaty" taste.

MAKING YOGURT AND CHEESE

The most successful byproducts from goat milk for the homesteader with just a few goats to make are yogurt and soft cheeses.

To make yogurt, first pasteurize the milk. Heat it to at least 162°F (72°C) and maintain this temperature for 15 seconds, before cooling it quickly to a temperature below 55°F (12°C). Alternatively, heat it gently to a temperature between 140° to 170°F (60° to 76°C), maintain this for 30 minutes, and then quickly cool it to 55°F (12°C) or less.

After this a yogurt culture must be added; special goat cultures are available and they will carry instructions for use. Alternatively, use store-bought plain live yogurt. Just add between a quarter- and a half-carton of plain yogurt to 1 pint (575 ml) of milk, then leave it covered in a warm place. A kitchen

Soft goat cheese is a simple and delicious way to enjoy your goat milk and it will stay fresh for longer than the milk itself.

cupboard is not really warm enough, but you could use a thermos placed in warm water or a special yogurt-making machine.

Leave the yogurt for up to about eight hours, then put it in the fridge. It will thicken as it cools but goat milk yogurt generally has a thinner consistency than commercial yogurt or that made with cow milk. You can use the yogurt you make as the culture for more yogurt for about a week. After that, it will begin to spoil.

Goat cheese

Both soft and hard cheeses can be made from goat milk, but soft cheeses are more suitable for most home production and can be made without any special equipment that you do not already have at home.

The problem with making a lot of cheese is that you can end up with a great deal of whey, which does not have much use. You could use it to try to make a sort of mysost, or Norwegian whey cheese. This is a thick-textured, light-brown cheese that has a very distinctive, caramel-like taste.

Strain the whey into a clean pan and bring it to boiling, stirring constantly. Skim off and retain the thick matter that comes to the surface and coagulates, then continue boiling and stirring the mixture. When it has reduced in volume by about three-quarters, return the coagulated matter to the pan and boil it some more, stirring vigorously.

When the cheese has thickened and is a light-brown color, remove it from the heat, but continue beating it until it is cool and too thick to stir any more. Pour it into a greased container and leave it to cool completely and set.

MAKING GOAT CHEESE

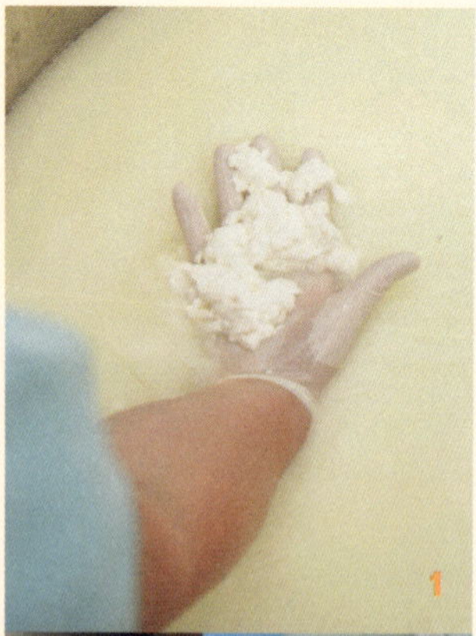

There are three easy ways of making soft cheese:

1 Allow the milk to sour naturally by letting it stand in a fairly warm place for 36–48 hours. Then wrap it in thick cheesecloth and let it drip into a bowl. After another 36–48 hours, you will be left with the cheese in the cloth and a bowl of whey beneath. You can also do this using yogurt you have made, but it will need to drip for about two days and has a rather sour taste. Try mixing it with garlic or chopped herbs.

2 You can also strain the cheese through perforated molds and then pour out the cheese, ready-formed into neat rounds.

3 In this method, rennet is added to the milk. Rennet is obtainable from health food stores, but make sure it is cheesemaking rennet. The rennet is likely to come with instructions for use, but the procedure is to warm the milk to about 90°F (32°C), add the rennet, let the mixture stand for about 30 minutes, and then put it through the cheesecloth. It will be ready much quicker than the other two cheeses and has a very mild taste. As it is the quickest cheese to make, the milk has less chance to assume a goaty taste. Shape the cheese into rolls or roundels, if you wish.

Preserving Meat and Fish

If you kill a chicken, it will probably only be one at a time and you can eat it right away, but if you slaughter a pig or catch a good haul of fish, you will need to preserve the results in some way to eat them over a longer period.

Home Smoking

Fish and meat can both be smoked, a process that adds a delicious flavor, even though it is no longer considered a reliable preservation method. If you have an open fire or a grill, you can smoke small quantities there, but a better solution is to turn an old wooden barrel into a smoker (or you can choose among several different kinds of commercial smokers). You need some space at the bottom for the fuel, which creates the smoke, and racks or rods above this for suspending the meat or fish. Finally, there need to be some holes in the side and top for the smoke to escape.

The temperature should not be high enough to cook the flesh, so allow the flames to die down until the fire is just smoldering smokily.

The wood or chips you use will affect the flavor. Pine is not recommended, as it will impart a hint of floor cleaner to your food. Better choices are oak, hickory, cedar, or wood from fruit trees.

If you can keep the fire at a consistently low heat, 10–12 hours should be long enough for smoking fish; bacon or hams will need a whole day or more, depending on their size and how strong you want the flavor. If you need to keep taking the meat or fish out while the fire burns too brightly, you will need to allow considerably longer—keep count of the number of hours that the food is actually in the smoker and work up to the total it needs.

You can smoke whole fish, fillets, or steaks of meaty fish, such as salmon. Experienced fish smokers prefer to brine the salmon before smoking it.

Freezing is the simplest, safest, and most effective way to keep produce fresh. For a large animal, a processor may also joint and cut up the carcass for you, or you could ask a local butcher to do it, perhaps in return for taking a cut or two.

Smoking and curing are other methods of preserving meat and fish and with the right equipment, you can do this at home. However, they will not preserve it for long, and are best considered as ways of adding flavor, rather than longevity.

Preparing fish

If you plan to freeze any fish you catch, put them on ice as soon as you catch them, keeping them in a cool box at the water's edge. When you get home, gut them, clean them, and scale them as soon as possible.

Dip each fish in very cold, salted water and allow it to drain well before wrapping in wax paper and foil and freezing. Rub individual fillets with olive oil to prevent them from drying out, then wrap separately or interleaved between sheets of wax paper. Fish will keep for six months in the freezer.

To scale fish, cover all nearby surfaces with newspaper—the scales will fly everywhere. Use a blunt knife to scrape away the scales, working from the tail towards the head of the fish. Take care not to tear the skin by holding your knife at the wrong angle or by pressing too hard. Rinse the fish under cold, running water to wash away any loose scales that are clinging to the skin.

To gut a round fish, insert a sharp knife into the belly of the fish at the vent and split it up to the head. Scrape out the intestines and other innards, taking care not to damage the flesh. Cut off the tail, head, and fins and open out the fish, pressing it flat on a board, with the skin side up. Press hard along the backbone to break and loosen it, then turn the fish over and gently and carefully ease the backbone and side bones away from the flesh with a knife. Check for stray bones that remain and remove them.

WATER AND ENERGY CONSERVATION

The Self-Sufficient Home
Conserving Energy
Keeping in the Heat
Solar Power
Alternative Energy Sources
Saving and Recycling Water

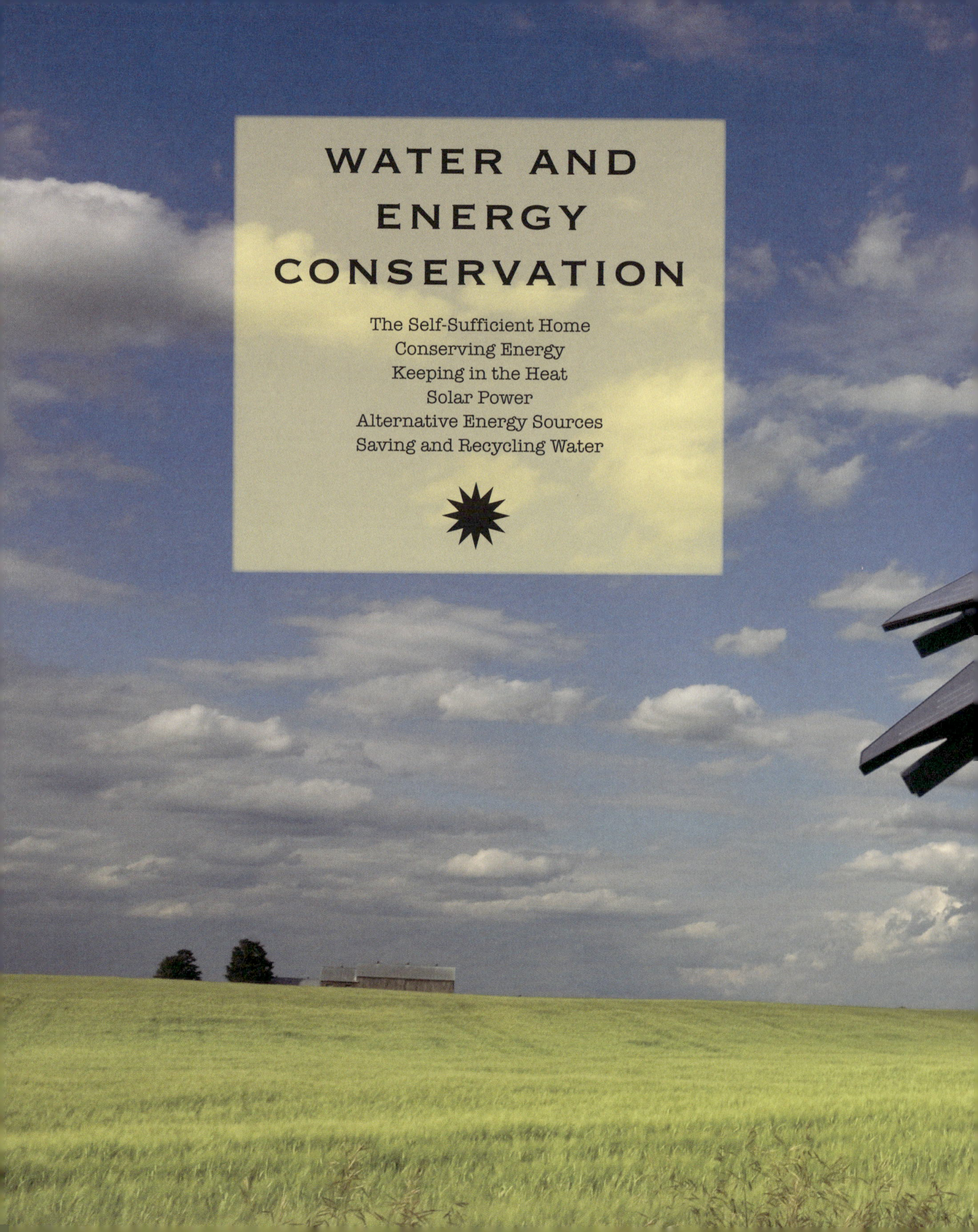

The Self-Sufficient Home

Being truly self-sufficient for your domestic water and energy supplies, as well as your food, is likely to be an impossible dream unless you have your own clear water supply, and even then it is not practical for most households. But you can get a long way towards this ideal of self-sufficiency by reducing your reliance on municipal water and power to the absolute minimum.

Packaging on prepared foods contributes in a large way to the problem of landfill. You can help by growing your own food, so that you buy less in the stores, but also by choosing products that are not over-packaged in plastic.

For many people, the driving force behind growing your own fruit and vegetables and keeping your own livestock for meat is very often a desire to minimize your impact on the planet, cutting down on unnecessary food miles, energy wasted by large corporations in producing, packaging, and selling food, the widespread spraying of chemicals onto commercially produced crops, and more.

Trying your best to reduce your own waste that goes into the garbage, recycling, composting, conserving energy and water, and maybe even generating some of your own power all go hand in hand with these aims and the ideal of self-sufficiency.

Tackling waste

The ultimate goal of the most enthusiastic self-sufficient homesteaders, and even urban and suburban homeowners, is to throw nothing away, using only things that can be reused or recycled, or that are packaged in recyclable materials.

This is unlikely to be practical in most cases, but whatever you can do to reduce the contents of your garbage cans will help to relieve the pressure on landfills and minimize the harmful gases and waste created as the garbage rots.

Try to support local suppliers and choose products that use as little non-recyclable packaging as possible. Read the labels: even some plasticlike materials can now be composted. And always think before you buy, to be sure that you really do need what you have picked out.

Saving energy in the home

Switching off lights when you leave a room and turning off any appliances or televisions that are left on in an empty room—these are obvious but nonetheless effective ways to reduce your energy consumption. Read the tips on the following pages to find out other simple ways to cut down on the gas and electricity you use without even realizing.

If you are lucky enough to have a large area of sustainable woodland on your property, it is worth considering converting your heating and cooking appliances to run on solid fuel, supplied from your own trees. Burning wood may be considered polluting, but as long as you plant trees to replace those you burn, it may be less damaging to the environment than the process required to produce electricity by commercial power generators, or the gas or other fuel extracted from natural gas or oil wells.

Being waterwise

Every drop of water that we use has been processed and treated to a high standard, even if we are only using it to flush the toilet or water the lawn. This is an expensive process, but also a wasteful one, using a great deal of energy and resources.

There are many ways to reduce your household water consumption (*see* pages 247-8) and the volume of water you flush away each year, relieving pressure on the water supply as well as saving yourself money. Always try to collect as much rainwater as you can. Not only is it free, but it is actually far better for your plants than watering them with highly chlorinated and treated tap water. Consider reusing your gray-water, too; that is, water used in the house for dishwashing, laundry, showering, and baths—this will not be potable water, but is often perfectly good for keeping the garden well irrigated.

Conserving Energy

Follow these simple tips in and around the home to reduce your energy use and help to conserve precious resources. You could be surprised at how much money you can save, too.

Get into the habit of always switching electrical appliances off properly. Consistently leaving your appliances on standby can add up to a hefty percentage of your electricity bill total. Of course, turning something on and off when it is a frequently used appliance is likely to be irritating and something you won't stick to, but make it a household policy to turn everything off around the house at the end of each day.

The same goes for lights. There is no excuse for leaving lights burning in an empty room, except in a handful of cases for security. Only turn the lights on when you really need to. Don't automatically switch on all the lights in a room if a single table lamp is sufficient, and try to avoid children getting into the habit of needing a light on to get to sleep.

Low-energy options

Incandescent light bulbs are being replaced by low-energy compact fluorescent lights. If you still have old-style bulbs in your house or in outbuildings around your property, it is worth replacing them. The new energy-saving bulbs last on average between six and 15 times longer than a traditional incandescent bulb and use 80 percent less energy to run. Most will pay for themselves within a year and last far longer than that. Replacing a 100W standard bulb with the equivalent 20W low-energy CFL (compact fluorescent light) can save you a good deal of money over the course of a year.

Don't heat empty spaces

Just as it makes no sense to light a room with nobody in it, you can also conserve energy by not heating rooms in your house that are not used regularly, such as spare bedrooms or the garage. Turn off the heater and close the door until you are expecting to use the room. Check the room regularly and air it so that it does not get musty or damp. It will soon warm up again when you need to use it.

If you have sustainable woodland on your property, burning wood may be a better choice for heating than using gas or electricity.

ENERGY-WISE HEATING

In cooler climates, heating your home accounts for a good chunk of your yearly energy bills, as does the cost of heating water. In warmer climates, you are likely to spend at least as much on cooling your home with air conditioning. Up to half of the average household's energy costs go to heating and cooling.

The boiler is the heart of any gas or oil heating system, so make sure that it is as efficient as possible. Have it serviced once a year. If your boiler needs replacing, choose an Energy Star boiler. These may be expensive, but are so much more efficient than the alternatives that they will pay for themselves, and more.

Controlling the heat

Make sure that your central heating controls are as advanced as possible. Fit radiators with individual thermostats that switch on and off according to the temperature in the room. This is a far more efficient method than switching all the radiators on and off according to a single whole-house thermostat.

If the temperature in the room with the main thermostat rises enough to switch the heating off, but you find that it is still cold in the room you are using, you will probably turn up the heating for the entire house to compensate. With individual thermostats you can set the temperature for each room independently: you can keep it warmer in a living room, where you are likely to be sitting still, for example, than in a kitchen, where the oven and general activity will add to the heat.

The most sophisticated systems allow you to set different target temperatures according to the day of the week and time of day.

Turn it down

Turning your thermostat down by just one degree can save you as much as 10 percent of your heating bill, and you are unlikely even to notice. Lower the temperature on your hot-water thermostat a fraction, too.

CHOOSING HOUSE-HOLD APPLIANCES

Always seek out energy-saving appliances when you need a replacement. All are now rated according to their energy-efficiency with the most efficient receiving an Energy Star label. Check the Energy Star website for up-to-date information about almost all home appliances. You may also be able to get subsidies for switching energy-inefficient appliances.

It is just as important to make sure that you are buying an appliance that is appropriate for your needs as it is to follow the manufacturer's energy ratings. A high-rated washing machine is no good for a large family if it is also a small one—running a large-capacity machine every couple of days is probably still more efficient than a higher rated machine that is too small for your requirements and needs to be used every day.

If you intend to freeze a lot of your produce to see you through the winter, consider investing in a chest freezer, but be realistic about how big it needs to be. Keeping a half-empty freezer cold takes a lot of energy; far better to have a smaller freezer that is well-stocked and therefore more efficient.

Where you keep your freezer will also affect its efficiency. If it is in the house, where the temperature remains fairly constant, it should run at its maximum efficiency, but in an outbuilding the temperature might fluctuate more. Make sure that it does not get too hot in summer in an uninsulated shed or garage, as it will have to work much harder there.

Washing clothes at a lower temperature can save a noticeable amount of energy without compromising on cleaning.

Keeping in the Heat

If your home is not well insulated and sealed against drafts, your precious heating will simply escape to warm the air outside.

Check your attic to make sure that it is adequately insulated. Warm air will rise up and out of a poorly insulated roof.

Good household insulation is vital to keep in your heating, but will also help to keep your home cool in hot weather, saving energy that would otherwise be spent on air conditioning or fans. Most people assume that insulating the home means doing the "big ticket" jobs, like upgrading the attic insulation or installing cavity wall insulation. These are, of course, the jobs that will save you the most if your house does not currently have either, but there are many small things you can do that can also have a noticeable effect.

Stop the small gaps

Using well-fitting curtains and closing them can make a big difference, particularly in rooms with old, drafty windows. This is especially true in cold weather, when the temperatures outside are considerably lower than those in the house. Buy heavy-weight curtains that fit close to the wall and install a pelmet at the top.

Inspect your front and back doors, too. Fit escutcheon plates to any mortise lock key holes, with holes that go right through to the outside, and install a mail slot cover flap to seal that against drafts.

Fitting draft stoppers

Gaps around doors and windows are as bad as leaving the door or window open. Install well-fitting draft stoppers all round the house and you could reduce the heat lost from your property by as much as 25 percent in the coldest months.

Keeping water warm

Make sure that hot-water pipes are insulated—split foam tubes that slip onto the pipes are quick to fit, easy, and effective, but you can also use fiberglass or wool insulation, wrapping it around the pipes in overlapping turns. Pay particular attention to bends in the pipe and valves that break the pipe run to make sure that the whole length of pipe is insulated.

Look at your hot-water tank, too, and make sure that it is well insulated. The best type is a modern cylinder with hard insulation sprayed onto the tank itself—it will feel solid if you tap it. If you still have an old, uninsulated tank, wrap it in an insulating jacket and make sure that it is securely in place.

Making big improvements

If your attic insulation does not meet the current building regulations minimum requirements, it is worth upgrading by adding an extra layer of blanket insulation across the whole attic, if you can. You may be able to get a homeowner grant towards the cost, either from your utility or power company or your state or local government. Even so, this does not have to be an expensive job, and will certainly help to make big savings in the long term. Hot air rises, and it will soon leak out of a poorly insulated roof.

Installing cavity wall insulation also qualifies for a subsidy in many areas. Solid walls are much harder to insulate effectively without compromising on the size of the room by fitting battens to the wall on the inside and a false wall on top of that to create an internal air gap.

Double glazing

Good glass makes a big difference to the temperature inside your house. The best double glazing will retain heat well, but also reflect heat on sunny days, helping to keep rooms cool inside. Special coated glass called low-emission (or low-e) glass is available, and the best low-emission double-glazed units are filled with argon gas to improve the insulation even more.

Solar Power

How much use you can make of solar power depends on where you live and the climate. In many areas, domestic solar panels are seldom able to generate enough electricity to take you off the grid, certainly not consistently throughout the year, but solar-powered hot water systems are more viable.

For generating electricity, you will need photovoltaic (PV) solar panels. With new developments, these are getting more efficient all the time, and you no longer need to cover an entire roof with panels to generate enough electricity to power an appliance. Nonetheless, they are very expensive to buy and install, and are extremely unlikely to be able to generate enough electricity for you to disconnect your property entirely from the national electric grid.

The most modern systems do work even on overcast days, though, and if you make an effort to cut down drastically on the electricity you use unnecessarily in the house, generating your own electricity might soon become a more feasible option.

Is it worth it?

Before you take the plunge and invest in some solar panels, you must do your homework. Consider how much electricity your household uses, what the likely peak output is for the system you have in mind—remember that you will only achieve peak output in peak conditions—and calculate how long the payback time is likely to be. For maximum efficiency, panels should be mounted facing south—they are usually fitted on a south-facing side of your roof—but they work on daylight, not necessarily direct sunlight, so they will also generate power on even a north-facing slope, but not as much.

Planning permission

You may need a building permit to install a bank of solar panels on the roof of your house. Check with your local building department before you do the work.

Power without cables

Photovoltaic solar panels are an excellent choice for powering lights in barns, sheds, and outbuildings that are not connected to the house's electricity supply, and this can make shutting in the chickens, early morning milkings, and evening rounds of your livestock an easier task. You can even power fans or

Solar panels work best when they are installed at an angle of around 60 degrees, to catch the maximum benefit from the rays of the sun.

heaters for a greenhouse from a solar panel, or pumps for watering or extracting water from a stream or pond for other use. Make sure that the panels are situated in a bright, unshadowed place where they can work most efficiently. You can also link up batteries to store energy for times when the solar panels are not producing enough power, but be sure to include a controller to prevent the batteries from overcharging.

If you want to install a solar system like this, it is a good idea to discuss your requirements with an expert first. They will help you to estimate how much power you are likely to need, how large your batteries should be, and how best to set up your miniature stand-alone circuit. A local sustainable energy organization is a good place to start. Alternatively, you can buy simple kits for installation, and these are usually good enough to provide at least some power.

HEATING HOT WATER

Solar thermal systems very simply use the heat of the sun to heat water for domestic use, and they are very effective. As long as the sun shines, you should have plenty of hot water in your tank.

Just like photovoltaic panels, solar thermal units for heating water are usually mounted on the roof of a house. Water pipes zigzag over a panel backed with a black, heat-absorbing plate, and this is all encased in a glass-topped, insulated box to minimize heat loss. Different systems are available, with differing degrees of sophistication and insulation. The more you pay, the greater the efficiency of the system; try to calculate how long it will take to pay for itself in heating savings, and which is the most financially viable choice for you.

Whichever type of panel you choose, the water that flows through the pipes is not the actual water you will be using in the house, but is a closed system of water pipes that then flow through the house's hot-water cylinder, warming the water inside, just as the electric element of an immersion heater does. It is a good idea to retain an electric immersion heater as a back-up (you cannot switch the sun back on if you suddenly run out of hot water in the evening), but in most cases, it should be possible to heat all or most of your hot water in this way, even in winter.

Just as with solar power panels, these are a good option for providing hot water in outbuildings, such as a dairy, where it would be difficult or prohibitively expensive to link up with the main house water system.

The more panels you have, the more power you will generate. Try to calculate how much power you need, and install the necessary area of solar panels to match it.

Alternative Energy Sources

Wind turbines are still quite new in domestic settings, and are more likely to be effective in rural areas, where there are not so many buildings to disrupt the wind. Use batteries to store energy created on a windy day.

Depending on where you live, you may have even more options for generating some or all of your own electricity from the wind or nearby water in a stream.

Commercial wind farms undoubtedly have a big impact on both inland and coastal areas, and seem to divide people into those who appreciate their value in terms of energy conservation and others who see them as eyesores. On a domestic scale, though, if you live in a suitable situation, a single-home wind turbine can reduce your electricity bill.

In urban areas, air turbulence around buildings makes turbines inefficient and they are rarely successful investments. But in a rural setting, where you have more space between buildings, they can be a viable option. The wind turbine itself should be mounted around 30 ft (9 m) higher than any obstacle within 100 yards (100 m).

Roof-top models with small wingspans can generate fairly small amounts of electricity, but stand-alone models, up to 45 ft (15 m)

Selling your power

In some states you may find that if you generate more power than you actually need, you can export some back to the electrical grid through some system of renewable energy credits (RECs). This will help to speed up the payback time of your investment in the generating equipment. But to do this you will also need to have an agreement. Make enquiries with your public utility commission or state power company to find out whether this is feasible before you make any additional purchases.

across, mounted on masts or stands will be more effective. Just like solar panels, the energy generated on windy days can be stored in batteries for use when the wind has dropped.

In order to decide if a home wind turbine would be economically viable for your homestead, you will need to know your Average Mean Wind Speed (AMWS). This information is available through various online tools, such as a Wind Energy Atlas or wind prospecting tools. When you know your AMWS, you can calculate whether a home turbine will generate enough energy to make it worth your while. Do your homework and check with several different sources before you commit to a home wind turbine. As with solar power, there may also be state or federal tax incentives to help finance the cost or setup of such a system.

If it is feasible for your property, it is a good idea to combine your solar panels and wind turbines into a "dual-fuel" generating system. This will reduce the possibility of you finding yourself without power.

Energy from water

Few people are lucky enough to have a running stream passing through their property, but if you are one of the fortunate ones and your stream is sufficiently fast-flowing, you may be able to harness the power of the water with a mini water turbine or micro-hydro generator. Just like an old-fashioned water wheel, the moving current makes the wheel spin, and a generator attached to the wheel makes electricity as it turns.

Look online for a reputable alternative energy supplier to find out what kind of turbines are available and to help you calculate whether your water source can generate sufficient energy.

Saving and Recycling Water

Conserving water is not just about cutting your water bill. The energy used to treat water for municipal supplies is a huge drain on resources, and climate changes mean that even the water itself is in short supply at times.

In a fruit and vegetable garden, there are many things you can do to minimize the amount of clean, household water you need to use. Chief among these is to set up an automatic irrigation system, and to collect as much rainwater as possible.

Position rain barrels or large water tanks at as many downspouts as you can. Don't forget outbuildings and sheds with pitched roofs. You can also link several barrels in a series if you have room, as one barrel will fill up surprisingly quickly when it rains, and empty fast, too, when you start to use the water.

Make sure that tanks have a secure lid or are covered with strong wire mesh that is not easy to dislodge. This is essential to avoid the risk of children or animals falling in and drowning.

Using the stored water

Most barrels come with a tap for filling a watering can. Watering by hand in this way is ideal, as you also have chance to check your crops for ripeness or signs of disease as you go, but it may not be the most efficient use of your time, particularly

Collected rainwater is perfect for watering plants: it is slightly warm and full of nutrients rather than chemicals.

at busy times of year. Rather than going back to using the outside tap at times like this, you can buy submersible water barrel pumps. These run on electricity, so you will need a power source nearby, but you can attach your hose to the pump and use it to water the beds with collected rainwater.

Underground tanks

If you are doing any large construction projects or earth-moving when setting up your garden, it may be worth considering installing a large underground rainwater tank. This will enable you to store far more water than you could in a collection of water barrels, and the water can be pumped up to an outlet for use.

GRAYWATER

It is a more complicated system to install, but "graywater" drained from baths, showers, washing machines, and sinks can also be treated and reused in the garden for watering your plants.

Before you consider a graywater system, be aware that graywater reuse may not be permitted under a municipal water supply system or even a domestic well permit. Check with your local Cooperative Extension Office or your state, provincial, or county health department to determine if there are any graywater use restrictions in your area.

A graywater system must be installed by a professional. The water must be treated, usually with a combination of filters and biological cleaners, before use. You will also need power to drive the necessary pumps for moving water through the treatment process and pumping it to the surface for use.

BEING WATERWISE

There are also ways inside the house to cut down on the amount of water you use every day.

Flushing the toilet accounts for around one third of the water used in most households. If you have old toilets with large tanks, each flush will be using far more water than is really needed. You can put water-saving devices inside the tank to reduce the volume of water stored for each flush—even a brick will do the job. And try not to flush the toilet after each use unless you have to.

When the time comes to renew your toilet, look for a water-efficient model that allows you to choose between a half and full flush.

Washing and waste

Think about the water you are using while you wash. Turn off the tap while you brush your teeth or shampoo your hair in the shower, then turn it back on to rinse. Take showers rather than daily baths, although remember that a five-minute shower can use as much water as a soak in the bath.

Fitting water-saving showerheads can help to reduce your usage, and don't overfill sinks and baths, or leave them running unattended. (If you run too much hot water, you will then need to add more cold water before you can use it.)

Dripping taps are an obvious source of waste, so always fix drips promptly. If you need to replace a tap, consider a flow regulator or a tap aerator. They provide a balanced flow to the temperature you want, but some lever-operated taps offer a low-flow setting that you must override with a harder push to get a full jet of water.

Stopping leaks

A leaking underground pipe left unnoticed will waste many gallons of water. If you have a water meter, get to know your normal usage and keep an eye on the reading from time to time to make sure that it is not giving an unusually high reading that could indicate a leak.

Another way to check is to turn off all the water-using appliances in the house, take a meter reading, wait for an hour, and then take another. The reading should be the same, unless water is seeping out through a damaged pipe.

Look out, too, for new muddy patches around your plot or garden, or a drop in the pressure from taps. If you think you may have a leak, contact your water supply company, who will inspect the pipe with a camera to find the crack, if there is one. If the leak is within your property's boundary, fixing it may be your responsibility and should be done as soon as possible.

A water-saving insert allows you to choose between a half and full flush, and uses less water than a conventional flushing system.

Take showers rather than baths to use less water, and fit aerating showerheads to be even more efficient.

Choosing appliances

Modern washing machines and dishwashers are far more economical in their water usage than they used to be, and the most efficient can wash an entire load in less water than you would use when washing by hand. The least efficient dishwashers are likely to use around twice as much water as hand washing a load of dishes.

Whatever machine you have, always wait until it is full before switching it on, to avoid wasting water on a part-load and then needing to use it again later on. If your machine has an economy setting, use it, and only buy appliances that are an appropriate size for your family's needs. Don't buy a family-sized dishwasher if there are only two people living in your house.

Finally, consider a solar clothes dryer, otherwise known as a clothesline. Your clothes will last longer and smell better, too.

WASTEWATER AND DRAINAGE

You can help to relieve the pressure on sewage systems by diverting your wastewater or excess rainwater into septic tanks and may be able to make yourself self-sufficient and cut off from the municipal sewage systems.

Many rural properties feed wastewater into a septic tank, as it is not economical to run drainage pipes in remote areas. These are large tanks and also incorporate a wide area of septic or leach field, so they are disruptive to install, but could be an option if you are digging up ground for another reason. Septic tanks need regular maintenance and emptying, so ensure that there is access for a truck to get to the site where you are planning to put it.

Composting toilet

If you want to install a new bathroom on your property, it is worth considering this eco-friendly option. They are difficult to install in a house, but are an excellent option in outhouses. They are surprisingly odor-free and completely natural. The toilet must be positioned over one of two open chambers, and a handful of sawdust thrown down with each use. For maximum efficiency, urine should be separated from solid waste with a simple filter, but this is not essential. Use as little paper as possible and never dispose of anything that is not easily biodegradable down the toilet.

Self-contained and central units with a composter in the basement are available. To be sure you don't run afoul of local health department rules, buy only from a reputable supplier (an NSF [National Sanitation Federation] certification is one indication of reliability).

The septic or leach field adjacent to the septic tank will dispose of rainwater run-off that cannot otherwise be stored in tanks or water barrels. This consists of a large, deep hole filled with gravel and other freely draining material. The water is fed through a network of perforated pipes in gravel, and gradually seeps into the surrounding soil, slowly enough not to flood the ground in heavy rain.

Over time, pipes become clogged with debris and need to be cleared and replaced nearby, or filled with fresh drainage material. This is a job for a professional.

INDEX

Page numbers in **bold** refer to illustrations

A

B

C

D

E

F

G

H

I,J,K

L

M

N

O

P

Picture Credits

We would like to thank the following for the use of their pictures reproduced in this book:

Alamy
162 tl © David Page, r © Image Source, 169 br © Rachel Husband,180 © Chad Ehlers, 207 © Gary K Smith, 233 © Eric Warren, 248 © Justin Kase z06z

Alison Candlin 52, 113

GAP Photos
12 John Glover, 21, 29 t Clive Nichols, 34 br Jenny Lilly, 35 John Glover, 38 br Michael Howes, 45 Frederic Didillon, 61 tl Mark Winwood, tr Rice/Buckland, 63 t Juliette Wade, c Sarah Cuttle, b Fiona Lea, 98 tr Juliette Wade, c Michael Howes, 100 tr Tommy Tonsberg, c Jonathan Buckley, 104 c Friedrich Strauss, 111 tr Rice/Buckland, br Howard Rice, 114 bl Friedrich Strauss, 133 tr Mark Bolton

Garden Picture Library
60 Michael Howes (Ansell Howes Assoc), 112 cr Juliette H Wade (Espalier Media Ltd), 133 bl Photos Lamontagne, 135 cr Michael Howes (Ansell Howes Assoc), 208 Geoff Kidd

Getty
2-3 Roger Charity, 26 Alan Buckingham, 33 t Mike Harrington, 72 Alan Buckingham

iStockPhoto
4, 5, 6, 10-11, 18, 20 t, 22 tl, 24 t, 31, 33 b, 34 bl, 36, 37, 38 bl, 39, 42, 43, 44, 46-47, 49, 51, 53, 55, 56-57, 58, 68 t, 73, 74, 75 tl, 79, 81, 82, 83, 86 t, 87, 88 br, 89, 91, 92, 93, 101, 102, 103, 105, 110, 118, 119, 121, 122, 123, 124, 126, 128, 129, 131, 134, 136 l, 139 t (x3), 154, 155, 158, 159, 163, 164 r, 165 r, 167 tr, 168, 169 t, 170, 171, 172, 173, 174, 175, 176, 177, 178-179, 181, 182 b, 184 bl, 188 br, 189 b, 192, 196, 197, 199, 201, 202, 203, 230, 237

Shutterstock
5 b, tc, 7, 13, 14, 15, 19, 20 b, 28 tl, 30, 41, 48, 67, 108-109, 140-141, 142, 143, 145, 146, 147, 152-153, 161, 166, 167 bl, 182 t, 183, 185, 186, 187, 188 t, 190-191, 193 bl, 194 bl, 195, 209, 218, 238-239, 240, 241, 242, 243, 244, 244-245, 246, 247, 249, 250, 253, 254, 256

StockFood
5 bc, 132, 165 tl, 204-205, 206, 210, 214, 215, 217, 219 tr, 221 t, 223, 225, 228, 234, 235, 236

Louise Turpin 29 b, 40, 200

Decorative chapter opener and running head artworks by Louise Turpin.

Front cover Clockwise from top left: Getty/Foodcollection, Getty/Lee Avison, Getty/FhF Greenmedia, Getty/Matt Armendariz, iStockPhoto, Getty/Rachel Weil, Getty/Frank Greenaway, Getty/Dorling Kindersley
Spine iStockPhoto
Back cover (all Getty) l Chris Whitehead, r Judd Pilossof, b Roger Charity

Some of the material used in this book originally appeared in ***Practical Self Sufficiency***, and ***Perfect Preserves***, published by Quarto Publishing plc and Quintet Publishing Ltd.

Author and editor Alison Candlin
Art Editor Louise Turpin